THE RIDDLE OF MALNUTRITION

PERSPECTIVES ON GLOBAL HEALTH

Series editor: James L. A. Webb, Jr.

The History of Blood Transfusion in Sub-Saharan Africa, by William H. Schneider

Global Health in Africa: Historical Perspectives on Disease Control, edited by Tamara Giles-Vernick and James L. A. Webb, Jr.

Preaching Prevention: Born-Again Christianity and the Moral Politics of AIDS in Uganda, by Lydia Boyd

The Riddle of Malnutrition: The Long Arc of Biomedical and Public Health Interventions in Uganda, by Jennifer Tappan

THE RIDDLE OF MALNUTITION

The Long Arc of Biomedical and Public Health Interventions in Uganda

Jennifer Tappan

Ohio University Press
Athens, Ohio

Ohio University Press, Athens, Ohio 45701
ohioswallow.com

Printed in the United States of America
Ohio University Press books are printed on acid-free paper ∞ ™

27 26 25 24 23 22 21 20 19 18 17 5 4 3 2 1

Library of Congress Cataloging-in-Publication Data

Names: Tappan, Jennifer, author.
Title: The riddle of malnutrition : the long arc of biomedical and public health interventions in Uganda / by Jennifer Tappan.
Other titles: Perspectives on global health.
Description: Athens, Ohio : Ohio University Press, [2017] | Series: Perspectives on global health | Includes bibliographical references and index.
Identifiers: LCCN 2017006596| ISBN 9780821422458 (hc : alk. paper) | ISBN 9780821422465 (pb : alk. paper) | ISBN 9780821445914 (pdf)
Subjects: | MESH: Malnutrition—prevention & control | Infant | Child | Public Health | Preventive Health Services | Uganda
Classification: LCC RA645.N87 | NLM WS 115 | DDC 362.1963/90096761—dc23
LC record available at https://lccn.loc.gov/2017006596

Dedicated to
Wednesday, Oscar, and Sid

CONTENTS

List of Illustrations ix
Preface xi
Acknowledgments xvii

Introduction 1
ONE Diagnostic Uncertainty and Its Consequences 11
TWO Medicalizing Malnutrition 37
THREE The Miracle of *Kitobero* 68
FOUR In the Shadows of Structural Adjustment and HIV 111
Epilogue: Remedicalizing Malnutrition and the Plumpy'Nut Revolution 136

Notes 145
Glossary 181
Bibliography 183
Index 211

ILLUSTRATIONS

Figures

I.1. Marasmus 2
I.2. Kwashiorkor 3
1.1. "Bed for metabolic studies," c. 1952 29
2.1. Child treated for kwashiorkor at the MRC Infantile Malnutrition Unit, Mulago Hill 39
2.2. "Kwashiorkor in a 17 month old Ganda boy, showing syringe feeding . . . through a fine polythene tube." 40
2.3. Label for milk packets 52
2.4. Magalita at fifty-two weeks 56
2.5. Petero at thirty-four weeks 58
2.6. Waswa and Nakato at sixteen and a half months and their mother 59
3.1 and 3.2. Women delivering therapeutic formula to malnourished children at Mwanamugimu, c. 1965 75
3.3 and 3.4. Cooking demonstrations at Mwanamugimu 77
3.5. Mwanamugimu Nutrition Rehabilitation Unit as seen from original entrance 79
3.6. Staff of Mwanamugimu Nutrition Rehabilitation Unit, c. 1965 79
3.7 and 3.8. Measuring using a handful and informal discussions 86
3.9. June 1969 calendar 88
3.10 and 3.11. Mwanamugimu certificates 90
3.12. Weight chart 93
3.13. Mother with her daughter during a follow-up visit two years after admission 95
3.14. Mother with her severely malnourished daughter on admission 96
3.15. Mother with her daughter and second child after three weeks at the unit 97
3.16. Mother together with her subsequent children after five years 97
3.17 and 3.18. Protected springs pictured during construction and following completion, Luka Mukasa seated on right 101

3.19. *Tusitukirewamu* club 103
3.20 and 3.21. *Bwamaka bulungi* and the "simple frame shelter," Luteete Health Center in background 105
3.22. Demonstrating how to prepare *kitobero* in rural area 108
4.1. Deputy Minister of Health, Mr. S. W. Uringi, delivering a speech at the opening ceremony of the MRC Child Nutrition Unit's Expansion in 1969 115

Map

I.1. Uganda 9

PREFACE

In 2012, I returned to the East African country of Uganda to continue an investigation of past efforts to prevent a severe form of childhood malnutrition. My objective was to interview a new set of informants and follow up with the elderly women and men who had generously shared their time and memories with me in 2004. Even though nearly eight years had elapsed, both Nabanja Kaloli and Ephraim Musoke greeted me as an old friend. Musoke even skipped the customary handshake, welcomed me with a highly uncharacteristic embrace, and then held my hand through our entire visit. What was most striking was the number of young children in many of these households. Musoke, Kaloli, and others spoke to me about this either directly by telling me of their struggles to provide for the growing number of grandchildren in their care, or by joking that I should take this or that child with me. They were only half-joking. I was there to ask about severe acute childhood malnutrition. They politely answered my questions and provided the information I asked for, but they made sure I heard about the children orphaned by HIV/AIDS and how this was weighing on them in the final years of their lives. This misalignment of interests between foreign researchers and those on the ground is such a common critique of global health that it has become cliché. When I first interviewed Musoke in 2004 and asked him what people in this part of Uganda did when a child became malnourished, he responded almost in exasperation. He would teach the parents to prepare a special food that both alleviates and prevents malnutrition. His exasperation spoke to the obviousness of the matter. It spoke to the fact that severe malnutrition was not a major problem in his community anymore, I should be asking about other things. But how severe acute malnutrition went from a major concern to one that invited exasperation is also a story that needs to be told.

In 2003, when I first decided to visit the Luteete Health Center, approximately thirty miles north of Uganda's capital city, Kampala, and several miles off the main tarmac road, I did not expect to find anything, I did not anticipate that I would ever return, and I certainly did not contemplate making Luteete the primary field site for this study. The Luteete Health

Center was worth visiting, even if only once, because in the mid-1960s, a few years after Uganda achieved independence from British colonial rule, the health center became the first rural extension of Africa's first nutrition rehabilitation program. A year later Luteete also became an epicenter of the violence perpetrated by Uganda's first prime minister, and for this reason it seemed unreasonable to expect the program to have made a lasting impact in the region. The program, which continues to serve severely malnourished children from the Mulago medical complex in the Ugandan capital, has been known since the mid-1960s as Mwanamugimu, the first word in a Luganda proverb (*Mwanamugimu ava ku ngozi*) often translated as "A healthy child comes from a healthy mother." When I arrived at the Luteete Health Center and began inquiring about Mwanamugimu, I was repeatedly told that she was dead. After confirming that the problem was not one of translation and my fledgling facility with the Luganda language, I learned that one of the midwives who had spent much of her life working at the Luteete Health Center was known to the people who lived in this region as "Mwanamugimu." Florence Mukasa had been so devoted to preventing severe acute malnutrition in young children that she continued teaching parents the principles of the Mwanamugimu program until the year she died. According to the women and men who have shared their stories with me over the years since that first visit, Florence Mukasa, and the Mwanamugimu program for which she was known, have had a considerable impact on nutritional health and wellbeing in the region served by the Luteete Health Center.

The Mwanamugimu program was part of a long history of nutrition work in Uganda and tracing that history involved weaving together highly disparate bodies of evidence. Archival materials that typically form the backbone of historical analysis, including memoirs, reports, and other documents held in England, Uganda, and the United States, have been key to my understanding of this history and its significance. The personal papers of physicians involved in the Mwanamugimu program and its extension to the Luteete Health Center furnished invaluable information on the innovative public health approach and its initial evolution. This material is complemented by a vast array of scientific publications and global health reports. Like the colonial archives that must be read with an eye to the imperial imperatives of their production, articles published in medical and scientific journals emphasize methods and results that make them rich in details of specific procedures and findings, but poor sources of information on the highly situated and variable nature of biomedical research. Individual people, dates, and other contingent factors are explicitly absent in accounts that present data as conclusive and universal. Extracting evidence from such sources entails unearthing a human story that is intentionally left out. Reports published by international agencies like the World Health Organization (WHO) and the United Nations Children's Fund (UNICEF) also extrapolate data of universal application from the local specificity of medical work and require a methodological approach intent on reading the local back in. My methodology also involved remaining attentive to the different registers of scale on which the history of nutrition work in Uganda operated, from the local, to the colonial and later national, to the global, and back again.

Over the course of three separate visits to Uganda since 2002, I conducted over fifty interviews with two distinct groups of people, whose testimony figure in the following analysis in very different ways. Interviews with Ugandan and expatriate physicians and scientists identified based on their involvement in Mwanamugimu and related nutrition work provided vivid accounts of the research that made Uganda an important international center of nutritional science in the mid-twentieth century. Extensive interviews and conversations with now elderly women and men living in the area surrounding Luteete foreground local memories of the health center's expansion as part of the Mwanamugimu program, the postcolonial violence that blunted this public health initiative, and the ongoing importance that the program continues to have in their everyday lives. Those interviewed in and around Luteete fall into two categories: first, a number of the elderly women and men were identified through photographs and by other informants as instrumental to the rural incarnation of the program; but I also interviewed individuals who were randomly selected based on their willingness and availability, as my translators and I walked along the roads and paths weaving through Luteete and the neighboring villages. I understood all of my interviews to be marked by a set of intertwining factors, including my own position as a young, white, female researcher from the United States. My apparent youth and the fact that, in 2003 and 2004, I was a both a graduate student and married created unanticipated confusion for many of the elders I interviewed in Luteete and its environs. The idea of a historian interested in medical work was equally perplexing for a number of the biomedically trained personnel whose memories also helped me piece together this history.

Memories of Mwanamugimu were inevitably influenced by the intervening period of insecurity and violence, especially for those who lived near the Luteete Health Center where political upheaval and war disrupted their lives in two distinct periods since the program began. I therefore developed a methodological approach that considered the realities of their more recent experiences as a filter or litmus test of what mattered most. In the course of the interviews that I conducted in 2004, I used photographs documenting the program at Luteete as a mnemonic device to remind informants of the less viable and meaningful aspects of the program—aspects that had long faded from their memories. Other components of the program were, notably, both widely remembered and remained a part of the living memory and social practice within the surrounding community.[1] These aspects of the program were often discussed with very little prompting and without the need of photos to jog memories—they had become an ongoing part of daily life. Applying Megan Vaughan's concept of social practice as a form of living memory to infant feeding, water collection, and intergenerational knowledge transfer revealed that aspects of the Mwanamugimu program were not difficult to remember, because they were not yet resigned to the realm of memory. I interpreted these more readily discussed and more widely known components of the program as aspects of Mwanamugimu that had an ongoing impact in the health and wellbeing of children in this part of Uganda.

A brief part of the research conducted for this study involved ethnographic methods of participant observation. During both my preliminary research in Uganda in 2002 and the beginning of my year-long period of fieldwork in 2003 and 2004, I participated in cooking demonstrations and out-patient meetings at Mwanamugimu. Mothers and guardians of rehabilitating children, who had good reason to mock my lack of skill in peeling plantains, helped me learn firsthand how to prepare a nutritious local food mixture for young children. In the group discussions with those who came to the unit on an out-patient basis, I listened as mothers and guardians expressed a wide range of concerns relating to the nutritional health of young children. One discussion, for example, concerned the gastrointestinal illnesses that appeared to come from milk that was potentially diluted with water of questionable safety. In 2012, my interviews sought to both augment the evidence gathered years before and to ask a number of questions that emerged from the intervening period of reflection and analysis. Thus specific questions replaced the photographs that I initially used to spark conversations and rekindle memories. I gave a number of my informants copies of a group photo taken during the initial years of the Mwanamugimu program, in addition to the sugar, eggs, tea, salt, and other foods that served as parting gifts. Although some took a moment to recognize themselves in an image that was over forty years old, as soon as they did, this photograph became a gift that was clearly treasured. The analysis that follows paints an image of biomedical research and public health programming in Africa that may, at first, also be difficult to recognize. In the end, I hope, it will be illuminating and valuable to those with an interest in African and global health history, and in the future of public health programming in Africa and other regions of the world.

ACKNOWLEDGMENTS

There are many who have made my work on the history of severe acute malnutrition possible—too many to sufficiently acknowledge here. Given the international scope of the nutritional work conducted in Uganda over nearly a century and the region's colonial past, this project entailed multiple visits to Uganda and the United Kingdom and I owe a great deal to those who provided hospitality and assistance along the way. I am indebted to the many helpful archivists and librarians at the Ugandan National Archives; the National Archives of the UK; the Wellcome Library; the London School of Economics; the Bodleian Library of Commonwealth and African Studies at Rhodes House, Oxford; the Cadbury Research Library at the University of Birmingham; and the Rockefeller Foundation Archive Center in Tarrytown, New York. With institutional affiliation from the Makerere Institute of Social Research, I also conducted archival research in Uganda at the Albert Cook Medical Library as well as the libraries of Makerere University and Makerere's Child Health and Development Centre, and I owe particular gratitude to Jessica Jitta for allowing me to consult the resources held at the Child Health and Development Centre, and to the exceedingly accommodating librarians at Makerere University and especially the Albert Cook Library on Mulago Hill.

Although I collected most of the oral evidence for this study in Uganda, I was fortunate to also have an opportunity to interview a number of very generous people in England and Scotland, including Margaret Haswell, Paget Stanfield, Mike Church, and Elizabeth Bray. I will not soon forget Elizabeth Bray, the daughter of Hugh Trowell, who not only deposited, at both the Rhodes House and the Wellcome Libraries, an extensive interview she conducted with her father, but also spent an entire afternoon with me sharing her memories, as well her own work documenting her father's life, and a reprint of his pioneering text, *Kwashiorkor*. In the evening following an exceptionally long interview, Stanfield and Church allowed me to photograph and record the material in several boxes, brimming with notes, reports, correspondence, music, and, importantly, images documenting their work in Uganda and Africa's first nutrition rehabilitation program. Together with the memories that they generously shared over two full days of conversation, their personal papers allowed me to piece together the establishment and evolution of the nutrition rehabilitation program that is at the center of this study. Moreover, Stanfield has, over the years, continued to insightfully and patiently answer my many additional questions, and at times his ongoing correspondence and support have served as an inspiration to me and a reminder of the remarkable dedication to child health and wellbeing exhibited by people like Paget Stanfield and Mike Church, to whom this history must in part be dedicated.

Oral testimony recorded in interviews with biomedical personnel in Uganda and the United Kingdom as well as conversations with elderly residents in the region surrounding the Luteete Health Center furnish the human side of what would otherwise be limited to the dry technicalities of a biomedical history. Several physicians who made time in their busy schedules to answer my tedious questions deserve special mention, including Roger Whitehead, who spent hours, just before he left Makerere to return to England, making certain that I understood the politics of protein deficiency in the post–World War II period, and the complex relationship between the British Medical Research Council and the nutrition rehabilitation program in Uganda. Drs. Philipa Musoke and Louis Mugambe Muwazi did their best to relate their memories of their fathers and predecessors, Latimer Musoke and Eria Muwazi. I am also grateful to Professor Alexander Odonga, Drs. Josephine Namboze, Chris Ndugwa, and John Kakitahi for their willingness to discuss their personal histories of their medical work in Uganda. The director and staff of Mwanamugimu not only made certain that I felt welcome, but took the time to teach me the principles and allow me to observe the nutrition rehabilitation program in its present form. Jennifer Mugisha has, since the very first day that I visited Mwanamugimu, been a welcoming friend whose ongoing work to improve nutritional health in Uganda serves as a reminder that hope for the future lies within the able hands of skilled and dedicated Ugandans.

Among my greatest debts are those that I have incurred in the region surrounding the Luteete Health Center. It is not possible to fully acknowledge the remarkable hospitality of the many people in and around Luteete who invited me into their homes and with great patience answered my many questions. Among those who shared their memories with me, I was especially fortunate to have had the opportunity to meet and interview Florence and Wilson Kyaze, and Kasifa and Bumbakali Kyeyunne, who are no longer with us. I will never forget the many conversations that I had with Ephraim Musoke and how he embraced me when I last visited him and his wife Catherine in 2012. Nabanja Kololi, my "Mama Mukono," and her daughter Caroline Nalubega, like many of the women in and around Luteete, will serve as an inspiration for years to come. Nor would these interviews have been possible without the guidance and translation services provided by the medical officer in charge of the Luteete Health Center, Jackson Ssennoga, a local teacher, Jemba Enock Kalema, and Hajjarah Nambwayo. Hajjarah's mother, Fatuma, whose laughter and friendship will be missed until I am, one day, able to return, and Jackson's wife, Sarah, will always remain dear friends. Finally I must also thank the primary midwife at the Luteete Health Centre, Susan, and the Community Health Worker, Stephen Maseruka Mulindwa or "Ssalongo," for their seemingly infinite hospitality and kindness.

The financial support that has made this project possible includes several traveling fellowships awarded by Columbia and Portland State University. I also received support during the initial writing process from Columbia University's Institute for Social and Economic Research and Policy, now known as the Interdisciplinary Center for Innovative Theory and Empirics (INCITE). Finally, an American Council of Learned Societies/Social Science Research Council/National Endowment for the Humanities (ACLS/SSRC/NEH) International Area Studies Fellowship provided the support needed take a leave from my teaching responsibilities in order to analyze additional evidence and thereby significantly revise and extend the project. I am also grateful to the Friends of History at Portland State University for their support, specifically in the production of the map situating Uganda, Buganda, and Luteete. Although the flaws are clearly my own, my work has also been significantly influenced by conversations with and encouragement from Marcia Wright, Greg Mann, Nancy Leys Stepan, Tamara Giles-Vernick, Holly Hanson, Sheryl McCurdy, Carol Summers, Neil Kodesh, Rhiannon Stephens, Barbara Cooper, Cynthia Brantley, Alicia Decker, Brandon County, Wendy Urban-Mead, Mari Webel, and especially in recent years, Melissa Graboyes, among many others. I have had the distinct honor of recently working with many insightful and generous students who have all influenced my thinking, but Cathy Valentine, Emily Kamm, and Jessica Gaudette-Reed deserve special mention. Over the years, my work has also been significantly influenced by one of the most generous scholars that I have been fortunate enough to meet, and I doubt that Jim Webb will ever fully realize the extent to which this and my future scholarship are indebted to him and his support.

The greatest sacrifices have been made by my family. I began this project before my daughter, Wednesday, and my son, Oscar, were born. It is now difficult for me to imagine what this study would have been without them in my life, except to say that the process may have been somewhat more efficient. Both are too young to fully appreciate why this work is so important to me, but they have grown accustomed to many nights and weekends with Mom away at a conference or working at the office. One day I hope that they know how much they are a part of what I do, even when it takes me away from them. Ultimately, I think their lives are even more enriched than they might otherwise have been. What I have asked of my husband and dearest friend, Sid, is more than I even care to admit, and he may never know how much I appreciate all that he has done, as an incredible father and supportive partner, to make this work possible. I am reminded on a daily basis of how fortunate I am to live in a time when such true partnerships are an accepted part of love and marriage and it is in this spirit that I also dedicate this work to my companion in life and in the field.

INTRODUCTION

The riddle of malnutrition, which proved puzzling to health workers in the East African country of Uganda, as in other world regions, concerned the syndrome now known as severe acute malnutrition. Severe acute malnutrition is the most serious and most fatal form of childhood malnutrition. Global estimates in the early twenty-first century indicate that the condition annually affects between ten and nineteen million children, with over five hundred thousand dying before they reach their fifth birthday.[1] The condition was first recognized, as a form of protein deficiency known as kwashiorkor, in the mid-twentieth century and for a time was a central international concern.[2] Severe acute malnutrition is currently defined in fairly simple terms, but is far from a simple condition. Children who exhibit "severe wasting" or a weight-for-height ratio that is less than 70 percent of the average for their age are seen to be suffering from severe acute malnutrition. Alternative markers include nutritional edema or very low mid-upper arm circumference measurements. Children diagnosed as severely malnourished require immediate therapy and run a very high risk of succumbing to the condition.[3] What is more, recent investigations suggest that even those who do survive appear to suffer from long-term impacts on their overall growth and development.[4]

Until the late twentieth century, the condition now diagnosed as severe acute malnutrition, or SAM, was thought to be two entirely separate syndromes. Kwashiorkor and marasmus, which are now recognized as extreme manifestations of the same condition, occupy opposing ends along a spectrum of severe malnutrition. Marasmus is defined as undernutrition or frank starvation with the extreme and highly visible wasting of both muscle and fat (see fig I.1). Kwashiorkor, on the other hand, is seen as a form of malnutrition and although the specific cause or set of causal factors that lead to kwashiorkor remain uncertain, kwashiorkor came to be associated with a diet deficient in protein.[5] In sharp contrast with the very thin appearance of children suffering from marasmus, the most important and consistent symptom of kwashiorkor is edema, or an accumulation of fluid in the tissues, which gives severely malnourished children a swollen and plump, rather

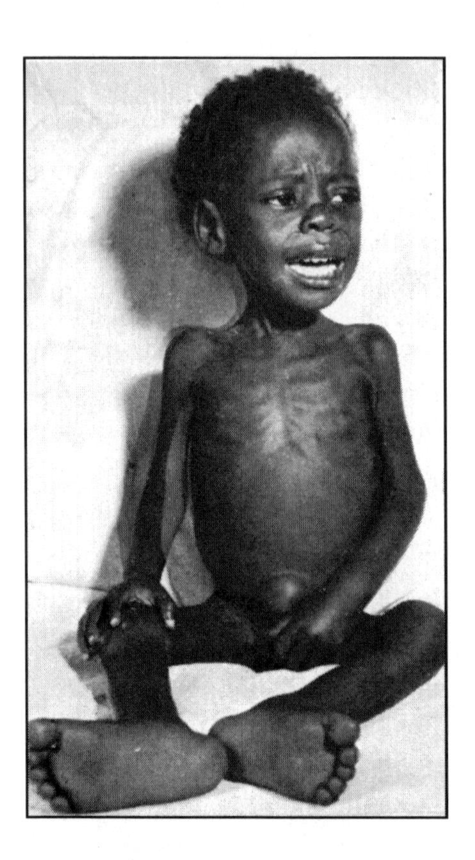

FIGURE I.1. Marasmus. *Source:* D. B. Jelliffe and R. F. A. Dean, "Protein-Calorie Malnutrition in Early Childhood (Practical Notes)," *Journal of Tropical Pediatrics,* December 1959, 96–106, by permission of Oxford University Press.

than starving, appearance (see fig I.2). This telltale swelling is exacerbated by an extensive fatty buildup beneath the skin and in the liver, and these symptoms long confounded biomedical efforts to understand the condition and connect it to poor nutritional health. Many children with kwashiorkor also develop a form of dermatosis, or rash, in which the skin simply peels away, and they often lose the pigment in their hair. One of the most distressing aspects of the condition is the extent to which children with kwashiorkor suffer. They are visibly miserable, apathetic, and anorexic.

The refusal to eat further exacerbates and contributes to an impaired ability to digest food, increasing the high mortality rates associated with severe acute malnutrition, and especially children suffering from kwashiorkor. Prior to the 1950s, case fatality rates in Africa ranged widely but were

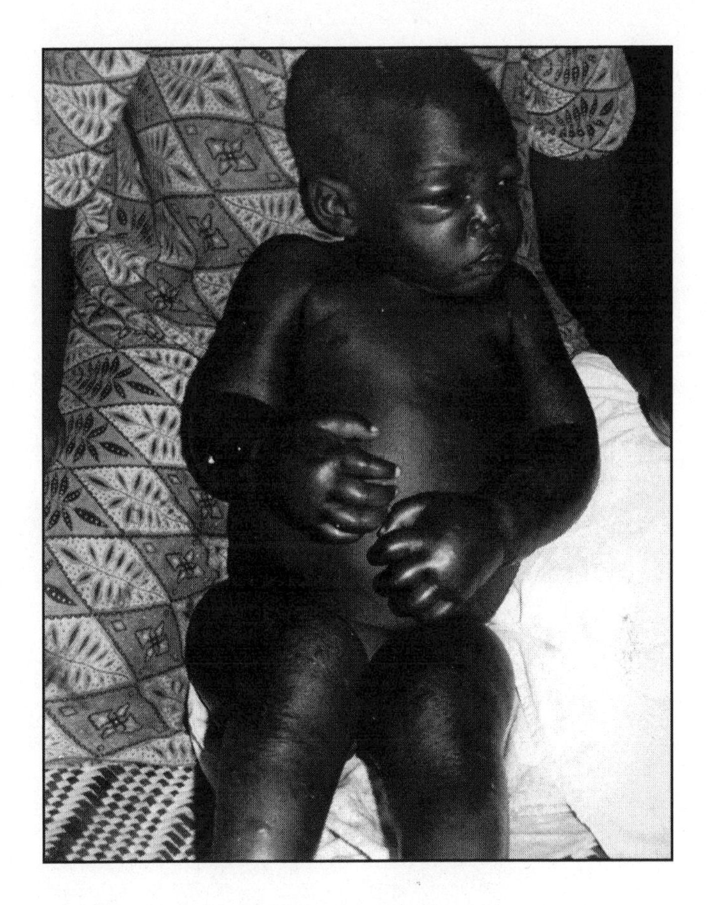

FIGURE I.2. Kwashiorkor. Courtesy of Paget Stanfield.

frequently cited as high as 75 and even 90 percent. Although these mortality rates fell considerably when new therapies were developed, they remained at unacceptably high global rates of between 20 and 30 percent until the twenty-first century. A new set of therapeutic protocols promises to reduce the associated mortality by more than half, but has only inconsistently achieved such rates of recovery and survival. Severely malnourished children who are infected with HIV experience the highest case fatality rates, which may have been a factor contributing to the mortality associated with the condition long before the discovery of HIV in the 1980s.[6] Moreover, severe acute malnutrition, like undernutrition more broadly, cannot be entirely separated from other forms of debility and disease, as poor nutritional health significantly increases the morbidity and mortality of a wide range of infections including HIV. Despite the "synergistic association" between undernutrition and disease, poor nutritional health is considered the cause of death only when recovery and survival are specifically compromised by the presence of malnutrition. On this basis, global estimates indicate that, taken together, various forms of undernutrition accounted for over three million deaths in children under the age of five in 2011—a figure encompassing an astonishing 45 percent of worldwide infant and child mortality.[7] The relatively small fraction of under-five mortality that is directly attributed to severe acute malnutrition alone—estimated in 2011 to be approximately 7.4 percent—nonetheless represents more than five hundred thousand children, a death toll on par with malaria.[8]

Like malaria, the prevalence of severe acute malnutrition is concentrated in particular world regions.[9] The overall global prevalence was estimated in 2011 to be approximately 3 percent, which roughly equates to nineteen million children, with the highest prevalence rates found in central Africa where an estimated 5.6 percent are severely malnourished. Global indications suggest that the prevalence of less severe forms of childhood malnutrition decreased since the 1990s, with Africa as the only exception.[10] Evidence from Uganda, where a 2010 survey found that 2 percent of children were severely malnourished and 6 percent showed signs of less severe wasting, corroborates this trend.[11] In the mid-twentieth century, annual prevalence in Uganda was estimated at 1 percent, although statistics from the Ugandan Ministry of Health indicate that, based on the twenty-three thousand to thirty-six thousand children annually diagnosed with malnutrition between 1961 and 1966, the prevalence may have been closer to between 2 and 3 percent.[12] Establishing even estimates of historical prevalence rates must confront a number of significant challenges. Prior to the 1950s, kwashiorkor was not widely recognized as a condition, and in Uganda, biomedical practitioners later equated a number of different locally recognized forms of illness as kwashiorkor.[13] What is more, prevalence has often been assumed to be more or less static, meaning that, until fairly recently and except when assessing the success of specific interventions, little effort was made to investigate the shifting epidemiology of severe acute malnutrition.[14] The result is that we are left with the knowledge that severe acute malnutrition remains a serious problem in many parts of the world, but have, at best, an incomplete understanding of how prevalence may have shifted over time. This gap in existing knowledge limits efforts to consider the role of contributing factors, including economic and social variables.[15]

In the postwar development era, when betterment schemes promised to lift entire populations out of poverty, severe acute malnutrition in Africa and other global regions did become a central international concern.[16] It not only occupied the attention of biomedical experts and nutritionists, but, due in large part to the jarring images of severely malnourished children that *Time Life Magazine* published in 1968—photos of children from the refugee camps of Nigeria's Biafra War—the condition also became the poster child for humanitarian aid and volunteerism. It became emblematic of one of the most pervasive tropes of the African continent: that of the starving African child. The Nigerian novelist Chimamanda Ngozi Adichie's poetic critique of the fleeting philanthropy that such imagery inspires—"Did you see? Did you feel sorry briefly…?"—speaks to the waxing and waning attention, and attendant funding, for such global health concerns.[17] The fluctuating international interest and investment in the problem of severe acute malnutrition reveal the role of biomedical research and programming in global health faddism. International interest in severe acute malnutrition has proven to be fleeting, and as global concern waned in the late twentieth century, past efforts to combat the condition were swept under the rug, all but forgotten. Galvanized by therapeutic innovations, a new generation of global health workers, biomedical experts, and philanthropists has succeeded in returning severe acute malnutrition to the international limelight. Without knowledge of prior initiatives, their work cannot build on previous endeavors or avoid past mistakes.

Enduring Engagements

One of the most important facets of nutritional research and programming in Uganda is that it represents a largely uninterrupted effort to understand and contend with a single condition for the better part of a century. This is remarkable because many, if not most, biomedical interventions in Africa from the colonial period onward have been time-limited. Public health projects and programs with end dates seek to make a lasting impact through temporary measures, and these measures are seen to either succeed or fail. Historians of colonial medicine and global health have, especially in recent years, drawn attention to the many insights that emerge from an examination of these targeted endeavors. A key observation is that Africans on the receiving end of health-related work are not merely passive recipients or biomedical subjects. Instead, they interpret and make meaning of health provision in ways that are tied to a complex set of local, and very often historical and thus dynamic, perceptions and experiences.[18] It has been shown that interpretations of both research and programming shape how people engage with biomedical projects and programs, how they are incorporated and thereby altered, or rejected and avoided.[19] The result has very often been a number of unintended consequences that typically impede stated objectives. Among scientists and public health experts, failure signals the need to return to the drawing board in order to devise better tools and methods, and in the global health context this typically involves packing up and returning to North Atlantic centers of research and policy development. Prior public health programming, especially when unsuccessful, is frequently then forgotten, leaving the lessons of past mistakes sadly out of reach for those launching at times nearly identical initiatives at a later date.[20] Yet, when these interventions come to the end of their time frame, they leave a trail of data and reports for later analysis and what historian Melissa Graboyes has recently referred to as "accumulated reflections." Interpretations and perceptions of medical endeavors *accumulate* among populations in Africa and elsewhere, and the residue of these past experiences continues to influence their view of and response to future efforts for perhaps decades if not generations to come.[21]

Recognizing that health-related work leaves a residue of past interactions and encounters and that even time-limited projects have a social afterlife illustrates one of the fundamental shortcomings of neglecting historical epidemiology.[22] Assuming that recipient populations are like blank slates in their perceptions of biomedical work overlooks how past experiences and programs influence future initiatives. While communities in Africa may at times be "biological blank slates," and thus represent unparalleled opportunities for testing new vaccines and drugs, their interpretation of and engagement with such work filters through the residue of past medical projects and programs.[23] The long history of nutrition research and programming in Uganda is an opportunity to examine how perceptions of biomedical research and treatment shaped the outcome of such endeavors. My analysis explores how the residue of past medical work fundamentally influenced how people engaged with biomedicine and public health. It asks how the residue of past experiences thereby influenced health-related research and programming. In tracing this influence it also reveals the dynamism inherent in local interpretations and interactions. Even "accumulated reflections" are open to change and, as nutritional work in Uganda shows, were highly responsive to the shifting research protocols and evolving programs of treatment and prevention that they themselves engendered.[24]

The nutritional work conducted in Uganda for the better part of a century challenges common definitions of global health. According to a recent volume titled *Global Health in Africa*, global health has its origins in colonial times, but emerged in the post–World War II period and can be defined "broadly to refer to the health initiatives launched within Africa by actors based outside of the continent."[25] Yet, in addition to the ways that local interactions influenced nutritional work in Uganda throughout the period under consideration, it is also true that Ugandans were pivotal to the nutrition research and especially the later programming that are at the center of this study. In fact, both biomedical training and infrastructure *in Uganda* proved crucial to the evolution and longevity of a public health program that continues to be of great significance to many Ugandans. The public health approach that emerged in Uganda was far more of a local endeavor than it was a global health initiative.[26] When initial efforts failed, biomedical personnel returned to the drawing board, but it was a drawing board *in Uganda.* The research protocols and initiatives that were then devised explicitly harnessed local engagement with biomedicine—they put enduring engagements in the service of public health. In fact, the period of failure that preceded the advent of this novel public health approach was one marked by Uganda's emergence as an international center of nutrition science and the post–World War II rise of global health.[27] In launching Africa's first nutrition rehabilitation program, the expatriate architects of the initiative saw the errors of existing global health models and devised a new one.

Mulago and the Kingdom of Buganda

The Mulago medical complex has long been the locus of biomedical research and provision in Uganda, and Mulago together with the region surrounding one rural health center constitute the two principal sites of fieldwork for this study. Mulago is one of the many hills that define the landscape of Uganda's contemporary capital city, Kampala. This urban center has also long been the political capital of Buganda, one of the numerous interlacustrine kingdoms that dominated this region of East-Central Africa prior to colonial imposition.[28] Stretching like a fertile crescent across the northwestern shores of Africa's largest lake, Buganda spans from the Nile River in the east to the Kagera River in the southwest. The kingdom's controversial northern boundary was extended under British suzerainty to the

Kafu River, finalizing a centuries-long process of territorial expansion and regional ascension (see map I.1).[29] European explorers and missionaries had been present in Buganda, alongside coastal merchants, since the mid-nineteenth century, but the British did not become actively involved in Buganda's political affairs until the late 1880s. Through amicable relations and a strategic alliance formalized with the British in 1894, Buganda and the port town of Entebbe became the political and economic headquarters of the British protectorate. Ganda participation in the pacification and administration of other areas within present-day Uganda placed Ganda in an advantageous position and created a divisive context with significant consequences following independence.[30]

The British were impressed by what they saw as an exceptional example of a progressive and sophisticated state in the heart of tropical Africa, and sought to govern indirectly through Buganda's highly centralized and bureaucratic structure of chiefs and royal officials. An agreement signed in 1900 established, for the *kabaka* (king) and the reigning Ganda chiefs, a degree of political autonomy and, notably, freehold rights to virtually all of the productive land in Buganda. Parceled out in estates so vast that they were measured in miles (and became known as *mailo*), land in Buganda was transformed into the private property of what then became an oligarchy of Ganda chiefs.[31] Even as it increased the power of chiefs vis-à-vis ordinary Ganda, this agreement kept Uganda from becoming a settler colony. Ugandans were able to thereby avoid the fate of those in neighboring Kenya and Southern Africa, where, by contrast, land alienation left Africans to subsist on the diminishing resources of overcrowded Native Reserves or as squatters and tenants on white-owned farms. The agreement, together with the completion of the Uganda Railway in 1901, set the stage for the rapid development of a flourishing export-oriented cash-crop economy, based initially on the small-scale peasant production of cotton and later on the far more lucrative cultivation of coffee. For average Ganda, most other avenues of upward mobility were effectively blocked by the Indian and expatriate monopoly on the processing and marketing sectors of the Ugandan economy.[32]

As the commercial and administrative center of the British Protectorate, Buganda was also the hub of both government and missionary education and medical provision. Albert Cook of the Church Missionary Society (CMS) established the largest and most successful medical mission station in East Africa on a hill not far from the capital or *kibuga* of Buganda. As in other regions of the continent, education and medical services for African

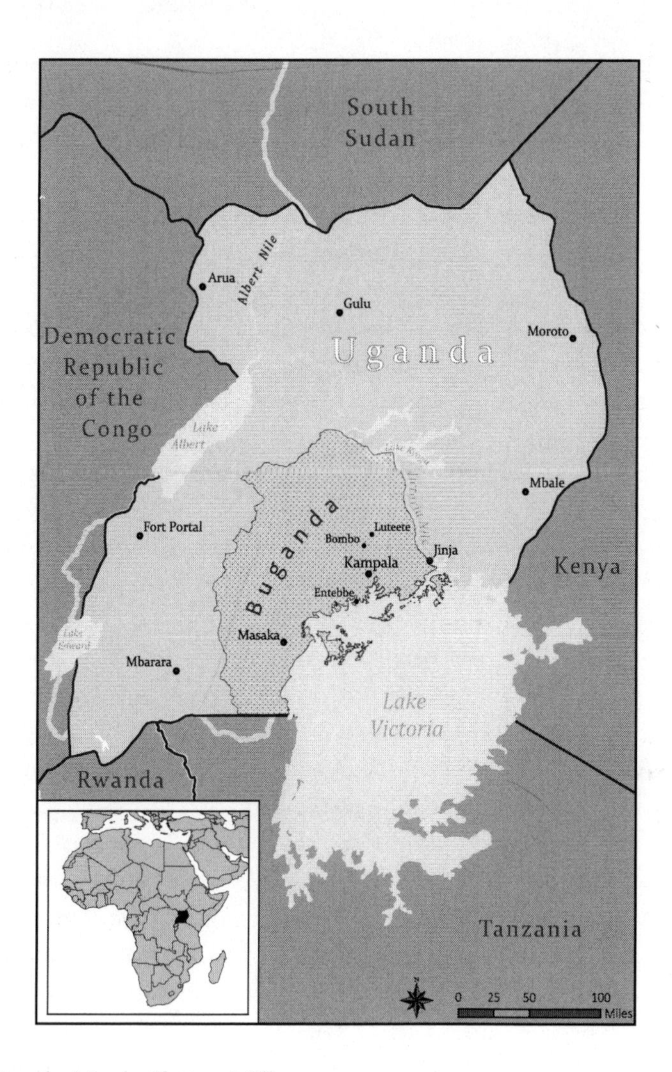

MAP I.1. Uganda. Map by Shawna Miller.

populations were initially the sole purview of the missionaries.[33] Particularly in the wake of a devastating sleeping sickness epidemic and concerns over demographic decline linked to venereal infections, the colonial government did eventually begin providing medical services and training medical auxiliaries.[34] The very high standards of medical training achieved at Mulago transformed the associated vocational school into a major research university. By the late colonial period, the Mulago medical complex was a center of research and training, drawing the best students from Uganda, Kenya, present-day Tanzania, and other regions of the continent.[35] It was this strong foundation of medical training that made Uganda a site of cutting-edge biomedical research—research initially focused on understanding and treating severe acute malnutrition.

The Riddle of Malnutrition traces longstanding efforts to understand, treat, and prevent severe acute malnutrition. These efforts initially served to medicalize the condition in the eyes of both biomedical personnel and the Ugandans who brought their severely malnourished children to the hospital for treatment and care. Medicalization meant that the condition came to be seen as a *disease* and a medical emergency.[36] My analysis explores how this understanding of the condition undermined prevention with unintended consequences, further imperiling the health and welfare of young children in Uganda. Biomedical personnel responded to the failure of prevention by launching Africa's first nutrition rehabilitation program. The program they designed aimed to demedicalize malnutrition, to learn from past mistakes, and it is one of the arguments of this book that the apparent efficacy and remarkable longevity of the nutrition rehabilitation program was the result of this critical reflection on the inadequacies of prior initiatives. Examining the perspective that was thereby gained reveals the immense value of historical epidemiology. It also shows how the advent of a novel public health approach to severe acute malnutrition built on Uganda's strong foundation of biomedical training and expertise and local engagements with biomedical treatment and care. As the program evolved it became a truly local initiative with a lasting legacy in at least one part of Uganda and with, at one time, aspirations to become a national program promoting nutritional health among all Ugandan children. How such a program could be largely forgotten outside Uganda is also a part of this history, and the potential implications of this unwitting amnesia are considered in a final examination of how recent innovations may return us to an earlier era when a medicalized approach compromised nutritional health in Uganda. This study is written in part to try to break this cycle of neglecting past public health initiatives as a new generation works to devise and advocate for policies, technologies, and programs that promise a healthier and more secure future for people around the world.

1 DIAGNOSTIC UNCERTAINTY AND ITS CONSEQUENCES

The early history of severe acute malnutrition is a history embroiled in controversy. Disputes over diagnosis plagued the condition from the very outset, and in Uganda these diagnostic debates stretch back to the turn of the century. In fact, consensus that the condition was a form of malnutrition did not emerge within the scientific community until the middle of the twentieth century. This lengthy period of diagnostic uncertainty was not without repercussions. As long as the etiology remained elusive, treatment was haphazard and largely unsuccessful. Ongoing efforts to determine the cause of the condition in order to devise an effective cure translated into years of experimentation on severely malnourished children, the vast majority of whom ultimately died. Insufficient caution or concern for how this period of diagnostic uncertainty might impact local views of biomedical research and care converged with mounting economic and political grievances, such that colonial authorities and biomedical personnel were finally forced to pay attention. The brief interruption in nutritional research that followed reveals "a nervous state," a colonial government responding to rumors and what they saw as superstitions in order to avoid further violence and unrest.[1] The resulting shift in research protocols furnish an opportunity to gauge how local engagement with biomedical research and care engendered changes that might otherwise remain obscure. Despite the advent of a more cautious approach and more effective forms of therapy, the consequences of this lengthy period of diagnostic uncertainty did not immediately dissipate, and local views of the nutritional work carried out in Uganda shaped therapeutic decisions with significant consequences for years to come.

"Groping Very Much in the Dark"

Diagnostic uncertainty surrounding severe acute malnutrition dates back to the early history of British colonial rule in Uganda and the early history of medical provision and training in the region. In the early 1900s, the preeminent medical missionary Albert Cook observed high rates of infant mortality and attributed the problem to congenital syphilis. Children suffering from syphilis, acquired during pregnancy and birth, exhibit a set of symptoms very similar to those with severe acute malnutrition, making a differential diagnosis difficult.[2] Both medical experts and historians have acknowledged that an inability to readily distinguish between different forms of syphilis and yaws contributed to the highly inflated and alarming prevalence rates cited in the early years of British colonial rule.[3] Few have noted that an unknown, but potentially significant, number of severely malnourished children were misdiagnosed as syphilitic in this period. Nor was this the only diagnostic dispute to stymie early efforts to diagnose and treat severe acute malnutrition.

Cook's view that venereal infections accounted for low birth rates in the protectorate was shared by his colleagues in government service. An investigation confirmed the exaggerated fears that venereal infections threatened demographic collapse in a region of increasing economic value to the British Empire, compelling the government to take immediate action.[4] The few treatment centers that were then built in and around Kampala aggravated an already existing shortage of medically trained personnel in the protectorate. The antisyphilis campaign thus gave rise to a modest training program, which became the foundation of the Makerere Medical School. Medical students at Makerere obtained their clinical experience in the wards of the central venereal disease clinic, turned general teaching hospital, on Mulago Hill. Before long the high standards of training achieved at Makerere made it the leading institution of higher learning in East Africa. It also attracted a new cadre of personnel, interested in both training and research.[5] The establishment of the Mulago-Makerere medical complex as the central medical institution in Uganda was so tied to this early antisyphilis work that the tendency to overdiagnose syphilis, and especially congenital syphilis in young children, continued in Uganda for much longer than might otherwise have been the case.[6]

Cook may have had the greatest stake in defending this diagnosis, as the colonial government also tapped Cook and his wife to start a maternity training school and establish a network of rural clinics as part of a further effort to reduce infant mortality and halt population decline. By the early 1930s, the Church Missionary Society (CMS) had already trained over one hundred Ugandan midwives and built more than twenty-five rural maternity and child welfare clinics, including the principal maternity center constructed on royal land in the village of Luteete.[7] Maternity training and the range of services Ugandan midwives provided for new and expectant mothers and their young children became central to the work and finances of the CMS medical mission station in Uganda. The fees charged at the maternity centers in 1931, for example, amounted to over 98 percent of the total annual expenditures, thereby subsidizing CMS work in Uganda beyond maternity training and provision.[8] There were therefore many who did not take kindly to suggestions that the condition most threatening the health and welfare of Ugandan children was something other than congenital syphilis.

The man who became the central figure in this diagnostic dispute was Hugh Trowell. Trowell first encountered children suffering from severe malnutrition in the early 1930s while stationed in the neighboring British colony of Kenya. Trowell was so inspired by the idealism of the interwar period and the creation of the League of Nations that he joined the colonial medical service immediately after completing his medical degree in the late 1920s. In Kenya, Trowell very quickly ran afoul of the settler politics limiting the provision of medical care, and he was transferred from his rural outpost to Nairobi where, in the context of the Great Depression, he expected to be dismissed from colonial service and sent home. A senior official who shared Trowell's interest in public health instead put Trowell in charge of a newly created African medical training program. Training African medical personnel beyond the level of hospital assistant did not have the support in Kenya that it did in Uganda. Reluctance to allow Trowell's medical students into the wards of the African hospital in order to obtain the necessary clinical experience forced Trowell to demand that he be given his own ward. When he first went to visit the pediatric department to which he'd been assigned, he found "about two children in each bed, and one underneath. Quite a number had brown hair. Some had swollen legs . . . and some were crying [and] moaning." Trowell later looked back on this moment as the first time that he came face to face with children exhibiting the classic symptoms of severe acute malnutrition.[9]

The consensus in Kenya was that these children were suffering from parasitic infections, especially hookworm. The problem with this diagnosis was that deworming medications did little to improve their chances of survival. Treatment failure was not only the linchpin in efforts to nail down the condition's etiology, but it also did little to encourage parents to bring their severely malnourished children to the hospital. Nor did it persuade parents to allow their children to be subjected to experimental procedures of limited apparent therapeutic value. In one of his earliest publications on the condition, Trowell reported that mothers were, as he put it, among "the greatest obstacles in the treatment. It proved necessary in these cases to separate completely the mother and child and to forbid suckling."[10] Only ten of the twenty-six children Trowell attempted to treat in this early study achieved a full recovery, and these were children brought to the hospital at an early stage of their illness. None of those who arrived severely malnourished survived. According to Trowell, five children "were discharged by impatient parents in an improved condition" but with an uncertain long-term prognosis, as "it usually proved impossible to detain them or secure their re-attendance."[11] Any resistance or reluctance on the part of parents is hardly surprising, given that Trowell later estimated he lost as many as three-quarters of his severely malnourished patients at this time. The largely futile attempts to treat the children in his care led Trowell to begin to suspect that he was dealing with a new disease.[12] Postmortem examinations conducted with the limited resources at his disposal revealed only that the children had an enlarged liver that was infiltrated with fat.[13]

A colonial medical officer stationed across the continent also became convinced that she was dealing with a new illness. Cicely Williams observed that the severely malnourished children at a children's hospital in present-day Ghana had been fed a diet deficient in protein and she was the first to propose that the condition was a form of severe malnutrition. Her seminal article on the condition documented how the provision of a milk-based, varied diet appeared to reverse many of the symptoms. In the end all but one of the children later died, but the visible improvement of their health in response to a high-protein diet led Williams to conclude that protein deficiency was to blame. Her hypothesis challenged the accepted diagnosis of experts in the burgeoning field of nutritional science. Hugh Stannus, who identified the condition as the vitamin B deficiency, pellagra, while working in present-day Malawi over a decade before, promptly published a review refuting Williams's evidence.[14] Williams responded with a second article in the preeminent British medical journal, the *Lancet*, delineating the significant distinctions between pellagra and the condition, which she then referred to as kwashiorkor, the local name for the condition in Ghana.[15]

The debate between Williams and Stannus reflected both the inconclusive therapeutic outcome of dietary treatments and the heightened interest in micronutrients following the wave of vitamin and mineral discoveries in the initial decades of the twentieth century.[16] Yet there was also a political dimension, as the proposition that children living within a British colony suffered from a diet deficient in one of the major food groups, rather than a newly discovered vitamin or mineral, was, in the words of a leading figure in British nutritional science, "politically objectionable."[17] Whereas previously unknown vitamin and mineral deficiencies provided further opportunities for science to improve the lives of colonial subjects, protein malnutrition pointed to the poverty of colonial populations. Williams was not easily deterred and despite her transfer to Malaysia (and interment during World War II), she remained an ardent advocate of the protein hypothesis throughout her life.[18]

The small medical library in Nairobi where Trowell lived and worked at the time did not carry a subscription to the *Archives of Disease in Childhood* that published Williams's first article advancing the protein hypothesis, and thus Trowell only learned of the debate through the refutation written by Stannus. Trowell then assembled a collection of photographs and tissue specimens and consulted with Stannus while on home leave in 1935. According to Trowell, Stannus only glanced over the photographs and then confidently reiterated that the condition was a form of pellagra. When pressed about the fat deposits in the liver, Stannus apparently refused to examine the liver specimens and sent Trowell away with an article on pellagra among African Americans.[19] For a time, Trowell nonetheless followed Stannus, referring to the condition as a form of "infantile pellagra" in his first two publications on the condition.[20]

In fact, Trowell did not begin to suspect that the condition was not a form of pellagra until relocating to the neighboring colonial territory of Uganda. He first traveled to Uganda in order to assess and report on medical training at Makerere. His enthusiastic endorsement of the high level of African medical training in Uganda was not well received in Kenya. Antipathy to advanced medical training for African colonial subjects was so great among the white settlers that Trowell's endorsement was met with a punitive relocation to a remote outpost in Kenya's Northern Frontier District. Only through an invitation to join the teaching staff at the Makerere Medical School was Trowell able to avoid being demoted and separated from his wife and young children. Once in Uganda, Trowell was again placed in charge of pediatrics where he again found children suffering from severe acute malnutrition. His further efforts to determine the cause and devise a cure placed him at odds with many within the medical establishment in

Uganda. The pathologist at Mulago Hospital, for instance, refused to thoroughly examine Trowell's patients at autopsy.[21] Albert Cook, who undoubtedly felt beleaguered by new legal restrictions on the services midwives could provide, which threatened CMS finances, was especially reticent to entertain the idea that the condition was not congenital syphilis.[22] In fact, Cook never conceded that children previously diagnosed with congenital syphilis may have been severely malnourished, and severely malnourished children continued to be treated as cases of syphilis in the main CMS hospital until the mid-twentieth century.[23]

Isolated, but undeterred, Trowell created his own small laboratory and trained a Ugandan assistant, John Kyobe, to examine liver extracts and blood slides. When a cable arrived from the United States announcing the synthesis of a new B vitamin and its arrival on the next plane, Trowell dropped everything to get to the airport, retrieve the bottle of niacin, and begin treating the children suffering from severe malnutrition in his ward.[24] Tragically, it was not known then that vitamin B therapies were harmful and dangerous to severely malnourished children, and eight of the ten children given the niacin died as a result.[25] Trowell resumed his efforts to treat the children with other B vitamins before eventually abandoning vitamin B deficiency as the cause of the condition, turning his attention to the role of anemia. In 1938, he used his time on home leave to return to England via Cairo in order to collaborate with an Egyptian doctor who had been publishing on anemia, and to take a three-month postgraduate course on anemia and other "blood diseases."[26] Once back in Uganda, Trowell enlisted Eria Muwazi, a recent graduate of the Makerere Medical School, and together Muwazi and Trowell conducted a number of studies in an effort to identify the cause of the condition and to assess its prevalence in the region.[27]

Muwazi's contribution to this research was significant. Muwazi, as a Muganda, could obtain detailed histories, dietary information, and more accurate appraisals of symptoms to compare with clinical examinations, biopsies, and blood tests.[28] Muwazi's involvement in the investigation of congenital syphilis and its true prevalence was pivotal, as he collected most of the data and ensured the success of the entire investigation, which found that only a very small percentage of children, perhaps as few as 1 percent, suffered from congenital syphilis. In fact, Trowell, Muwazi, and another researcher involved in the investigation concluded that congenital syphilis was uncommon in Buganda, particularly when compared to what they claimed was the "almost universal" prevalence of severe acute childhood malnutrition.[29]

It is not clear how they collected blood specimens in these preliminary studies, but the published findings indicate that in the early 1940s, they did not yet have the means to estimate blood protein levels in Uganda. Serum protein evaluations were therefore conducted on only a small fraction of the blood samples drawn for these early investigations and, as Trowell explained, "they used to take blood . . . and put it on a plane, and send it to Nairobi."[30] Over 180 severely malnourished children did, however, have blood drawn for a variety of other tests, including red blood cell counts and tests for anemia and congenital syphilis.[31] A subsequent study involving more than 120 Ganda children provides the first glimpse of what appears to be a new method of acquiring blood samples from young patients in Uganda—withdrawing it from a vein in the neck. This became a standard practice and remained the preferred method of obtaining blood specimens from young children for several decades. According to Muwazi and Trowell, the method was both efficient and effective: "In all children blood for serological examination was removed from the jugular vein in the out-patient department. This was done without much difficulty and it was found to be the method of choice as it required no special preparation. It was found that with a little practice 3–5 c.cs. of blood could easily be obtained."[32] This marked not only the inauguration of a new procedure, but an overall expansion in blood extraction on Mulago Hill. As Trowell later explained to his daughter, "At this time, I was working very much with the blood side of it. . . . But I was still feeling myself groping very much in the dark."[33]

In this early period of nutritional research in Uganda, severe acute malnutrition was not narrowly defined as a condition of early childhood and adult patients figured in a number of nutritional studies.[34] Adult subjects were important, as it was found that repeated blood extraction was dangerous in very young children.[35] The inclusion of adults also meant greater awareness of this work within the surrounding communities. Large numbers of immigrants from Rwanda and Burundi who came to work on Ganda farms through most of the colonial period arrived in Uganda so severely malnourished that they became the subject of numerous nutritional studies.[36] One study, for example, entailed extracting blood from 144 adult immigrants from Rwanda and Burundi.[37] Adults from the south-central Kingdom of Buganda were also included in investigations of childhood malnutrition as, for instance, in the 1947 investigation of 128 Ganda children that entailed extracting blood from their mothers for the very same blood tests.[38] Another study involved extracting blood from either an umbilical cord or a newborn's skull in order to compare it with venous blood taken from the newborn's mother.[39]

The establishment of a physiology and biochemistry laboratory on Mulago Hill in 1946 meant a further expansion of blood work in Uganda. The new laboratory equipment made more complex examinations of blood samples possible, thereby eliminating the need to send them to Nairobi. One researcher, Dr. Ferdie Lehmann, explicitly came to investigate anemia, and reports and publications reveal that blood was the central focus of the research conducted at the lab for a number of years.[40] Researchers working at the lab were able to conduct sizeable studies; thus, one investigation involved taking blood from 260 men chosen from among patients who sought treatment at Mulago's outpatient department.[41] Studies of blood protein levels involving such large cohorts of adults and children suggest that knowledge of this blood work could have spread in the area around Kampala, if not farther afield. The fact that Muwazi and Trowell did not confine their investigations to patients brought to the hospital must have increased this likelihood. In the post–World War II period, Muwazi and Trowell conducted a study at the Budo and Gayaza high schools—the two most prominent and well-known schools in Uganda. The students were divided into groups and each group was given a different meal before Muwazi and Trowell measured their blood pressure, weight, and height and took blood from each of the students.[42]

In the 1940s a growing number of scientists and physicians working at hospitals and clinics around the world also began to investigate and publish findings on the condition. Earlier publications and reports came from such far-flung regions and gave the condition such a wide variety of names that there was little awareness of their mutual interest or the global prevalence of the syndrome. With the "pellagra controversy," as it came to be known, the debate over the etiology became the subject of an international exchange in the journals of pediatric and tropical medicine. This allowed Trowell to compare his findings with research conducted in South America, the West Indies, Asia, and especially other regions of Africa. As the tide turned against pellagra, the predominant focus of this research was the only pathological anomaly routinely found at death—the fatty infiltration of the liver.

Trowell and his colleagues used biopsies or specimens taken from live patients as part of their efforts to understand these unusual fat deposits in the liver. As Trowell later described the procedure, it is clear that liver biopsies were dangerous and in at least one instance resulted in a patient's death: "We would put . . . [the patients] under an anaesthetic, and would put in a large bore needle, with which I could suck a small thread out. . . . I wasn't doing it with as good needles as they have now. We hadn't lost any children. We . . . had lost one adult over this . . . because I hadn't realized how deep you could go in on a thin patient. It had made him bleed, I am afraid fatally."[43] Biopsies causing even one patient to die could alone generate local concerns, yet when it came to biopsies performed on young children, witnesses may have also had reasons to conflate biopsies with blood work. Trowell began conducting liver biopsies on young children after acquiring a special bore needle from Joseph and Theodore Gilman during a visit to South Africa in 1947.[44] When Trowell demonstrated the procedure at a conference, he reported that in the fifty to sixty biopsies he had performed on severely malnourished children, "almost invariably he had found that the liver was being pushed forward and that blood collected in the syringe."[45] Thus, in this period, biopsies and blood extraction and the relative dangers of each procedure were, for all intents and purposes, largely indistinguishable.

Biopsies were also routinely performed alongside blood tests, further blurring the distinction. This was particularly true of the nitrogen balance studies carried out at the physiology and biochemistry lab. As nitrogen is a primary component of all proteins and a by-product of protein metabolism, measuring the amount of nitrogen consumed and then excreted was the most effective way of quantifying the amount of protein used by the body. Highlighting the extent of the extraction involved in these investigations, one nitrogen balance study entailed closely measuring the nitrogen consumed and excreted in addition to collecting liver biopsies, blood for red blood cell counts, and serum protein estimations, *repeatedly* throughout the duration of the study. In order to obtain the most accurate results, the nitrogen balance studies were extended until, as one researcher noted, "the patient had to be granted discharge, or in default of that, absconded, from hospital." This meant that in some cases the investigation lasted for an astonishing 170 days.[46] Exact figures for the number of participants were not provided, but one report indicated that "the limit indeed, was not the supply of cases, but the working capacity of the laboratory,"[47] suggesting that the number of people involved in this particular set of studies may have been considerable.

During the Second World War, Trowell also sent liver specimens to Professor Harold Himsworth at the University College in London.

Himsworth became interested in Trowell's work when, in the context of wartime rationing, he was asked to determine the dietary protein requirements of young children. As very little was then known about the repercussions of protein deficiency, Himsworth began conducting experiments with rats and found that, like Trowell's patients, when deprived of protein they developed a fatty liver, discolored hair, and even slight edema.[48] Himsworth then became a key advocate of Trowell's work. Trowell preserved the liver specimens from his severely malnourished patients in his home refrigerator until they could be transported to England: "I would keep the bit refrigerated until I could take it to the plane, in the hope that it hadn't 'gone bad', as you might say, by the time it got to London. I'd put some of it in pickle sometimes."[49] He would, of course, as he later explained, warn his wife by letting her know "look, you mustn't let the boys cook this."[50] What the "boys" who cooked for Trowell and his family thought of the pickled liver specimens that were kept in the home refrigerator will probably never be known. It is, nonetheless, likely that such practices spurred scandalous rumors about the colonial medical officer conducting research on severely malnourished children.

The breakthrough in the search for the etiology of the condition finally came at the end of the Second World War when a new pathologist, Jack Davies, joined the staff at Mulago. Davies was already aware of Trowell's nutritional research before he arrived in Uganda, and, unlike his predecessor, was eager to assist in future nutritional studies. He immediately agreed to conduct thorough postmortem examinations, and in the first child he examined, Davies found the long-awaited pathological link between the highly fatal condition and protein deficiency. The child's pancreas was atrophied and the degenerate condition of the pancreatic cells indicated that the child's pancreas had not been secreting digestive enzymes. This crucial discovery appeared to substantiate Williams's protein hypothesis, as enzyme synthesis is dependent on adequate supplies of dietary protein. It was then that, in order to honor Williams and her foresight, Trowell began to refer to the condition as kwashiorkor. Pancreatic atrophy also explained why severely malnourished children were not easily treated. Even protein-rich foods like milk were of limited therapeutic value without sufficient protein to produce the digestive enzymes needed to breakdown and absorb essential nutrients. It was a vicious cycle: children suffering from severe acute malnutrition simply could no longer fully digest and absorb the food they consumed.[51]

To verify that the pancreatic atrophy occurred prior to death and not as part of a process of rapid decomposition, it was necessary to perform further autopsies within twenty minutes of the child's death. This was not possible during the day when both Trowell and his colleagues were needed in the hospital wards. "It was possible at night if," as Davies explained, he and Trowell "coordinated well." Trowell would inform Davies that "he had a dying child expected to die in the night. . . [and Davies] would wait ready in the morgue . . . till a scuffle outside indicated Hugh's arrival with the body of the sad victim he had pronounced dead only a few minutes before."[52] So little time had passed between when the children were pronounced dead and their delivery to the morgue that according to Davies, "usually the muscle twitched as [he] rapidly did a postmortem getting the essential organs into fixative as the time dictated."[53] Trowell and his colleagues were aware of the sensitivity that such work required and did take precautions. They would only remain in the morgue long enough to get the specimens into the fixative, and would wait until the next morning when they were delivered to the laboratory to analyze them. "For," as Davies explained, "never were Europeans allowed by the staff to carry anything other than papers from the morgue. This was in deference to local feelings. . . . Autopsies performed by Europeans in Mulago Hospital were closely watched, as were the pathologists. It would have been a very upsetting thing if a new pathologist, lithe and slender had become stout for the darkest suspicions would be aroused."[54] In the end, as we will see, such precautions proved to be insufficient and did little to assuage a growing set of local concerns.

This important development in the long search for the etiology of kwashiorkor came at a significant moment in the rise of international medicine. In October 1949, an Expert Committee on Nutrition formed jointly by the United Nations Food and Agricultural Organization (FAO) and the World Health Organization (WHO) held its first meeting in Geneva. Kwashiorkor, described as "one of the most widespread nutritional disorders in tropical and sub-tropical areas," was high on the agenda, and Trowell was asked to prepare a memorandum on the condition for the committee's consideration.[55] The committee resolved to conduct an investigation of kwashiorkor in sub-Saharan Africa beginning with a visit to Mulago Hospital in order to first consult Trowell and his colleagues in Uganda. The subsequent WHO report, *Kwashiorkor in Africa*, was based in large part on evidence from Mulago and became the seminal study in the growing international focus on protein malnutrition.[56] A second meeting of the Joint FAO/WHO Expert

Committee on Nutrition centered largely on discussions of this seminal report, concluding with a resolution to conduct further surveys.[57] Delegations later sent to Central America and Brazil confirmed that kwashiorkor was a worldwide problem requiring immediate action.[58]

The pathological evidence that appeared to connect the condition to protein deficiency did not immediately gain widespread acceptance, however, especially in Uganda. The pancreatic atrophy explained why it was so difficult to treat severely malnourished children, but it did not provide a clear way to address this problem. Experiments with pounded steak and desiccated hog stomach were disappointing and, in Uganda, as Davies later explained, they simply "had no special high protein material to feed the children."[59] Ongoing therapeutic failure fueled doubts that were compounded when the pancreatic atrophy found at autopsy could not be independently verified. At the time it was not known that the changes to the pancreas were only visible in children who died prior to receiving any treatment. The rapid recovery of the pancreas in children given sufficient milk reversed the signs of pancreatic atrophy found by Davies and Trowell.[60] The next step was to examine the secretion of pancreatic enzymes, and the British Medical Research Council (MRC), now headed by Himsworth, sent an expert who had been working on a new procedure to extract the contents of the small intestine from a tube inserted through the stomach with the guidance of an x-ray. The procedure furnished further evidence that the production of digestive enzymes was severely suppressed in children suffering from severe malnutrition. But before they had a chance to complete the analysis of their findings and publish the results, nutritional research in Uganda was swept up in a political insurrection with far-reaching consequences for future medical research in the region. As it turns out, this further extraction of fluids and tissues from severely malnourished children, who had very slim chances of survival, may have actually done more harm than good.

Muwazi and the Insurrection of 1949

On April 25, 1949, thousands of Ugandans gathered at the central palace of the kabaka, the king of Buganda. By 10:00 a.m., an estimated four thousand people were reportedly pressing against the palace gates. Leaders of the political organizations representing the protesters were allowed to enter the palace and present the kabaka with a set of demands, but tensions remained high, and more than a thousand people reassembled at the kabaka's palace early the next morning. Attempts to disperse them with baton charges erupted in violence. Arson and looting spread rapidly throughout the Ugandan capital and into the outlying districts of the south-central Kingdom of Buganda and continued for several days.[61] The violence explicitly targeted the property of specific individuals, including members of the Ganda chiefly class, elite Ganda officials, the Indian community, and a Ugandan doctor, Eria Muwazi.[62] Muwazi was at the central Ugandan hospital on Mulago Hill when news broke that his property and home had been destroyed. Muwazi feared that he and his family were in imminent danger and unsuccessfully sought police protection.[63] He believed he was "a marked man" because, as another medical officer reported, "Muwazi is said to kill children by taking blood."[64]

Mounting dissatisfaction with the ongoing experimentation on severely malnourished children who continued to die converged, in the late 1940s, with political unrest, contributing to what was already an explosive situation. Attention to this moment in the history of nutrition research in Uganda sheds light on how parents and guardians of severely malnourished children and their communities viewed these initial efforts to determine the cause of the condition. Pickling liver specimens and keeping them in the refrigerator until they could be flown to Europe would spur rumor and raise suspicion in any context. When combined with the risky and often experimental procedures performed in this period, such practices were sufficient to raise local concerns. But what made these practices especially alarming was the fact that, prior to the early 1950s, little if any progress had been made in the treatment of severe acute malnutrition. As Trowell later acknowledged, they were "not getting much information out of it."[65] This meant that the biopsies, autopsies, nitrogen balance studies, and extensive blood taking had not yet translated into clear benefits for dying children. In the period leading up to the 1949 insurrection, mortality rates associated with the condition remained very high, with rates at Mulago still in the order of 40 to 60 percent and WHO citing mortality rates in Africa as high as 90 percent.[66] Against this background, accusations that Muwazi "kill[ed] children by taking blood" became emblematic of an escalating set of economic and political grievances in late colonial Uganda.[67]

Across the continent, the 1940s were a period of heightened political dissent and labor activism. In Uganda, workers calling for higher wages and crop prices organized a general labor strike in 1945, and in both the labor strike and the political insurrection that followed, political activists sent a clear message that the increasing economic constraints of the postwar period represented a failure of political leadership.[68] Despite distinct visions of how best to achieve a more just and equitable society, activists shared a strong objection to the unethical abuse of power for personal gain in colonial Uganda. The landed oligarchy of Ganda chiefs and Indian middlemen became obvious targets of unrest, as postwar inflation created great hardship for ordinary Ganda and intensified longstanding accusations of profiteering on the part of the Indian middlemen.[69] The worsening economic situation exposed the practice of indirect rule in Buganda as an inherently autocratic and oppressive system. The inaction of Ganda chiefs and officials in the face of economic and political threats to general welfare and wellbeing indicated that their alliances were with the British rather than with the people, making them targets of the insurrection.[70] When the delegates met with the kabaka in 1949, they sought to critique this enduring alliance between the British and the landed chiefly class, and they notably couched their grievances in references to hunger and starvation.[71] One representative told the kabaka, "people are undernourished, they eat bad food because they have no money." Another claimed that "the people outside there are in agony. . . . The growers are dying from hunger."[72]

Many of the grievances that fueled unrest in this period coalesced in the figure of the elite Ganda doctor and aspiring politician, Eria Muwazi. After graduating from Makerere in 1934 with the prestigious Owen Medal, Muwazi became the senior African medical officer and "medical tutor" or "African Registrar" at Mulago, making him an especially prominent member of the medical profession in Uganda. "By the later 1940s," the historian John Iliffe notes, "medical graduates had become the elite of the elite. . . . They were professional men . . . with growing families and many social contacts. They were invited to tea at Government House [and] became the first African members of official boards."[73] For Muwazi this was especially true. As the central figure in the formation of the Makerere Medical Graduates Association, Muwazi led the struggle for official recognition of their professional status. He was the first East African to publish in a scientific journal, later became a high-ranking politician closely connected to the kabaka, and was named the third most important figure in the parish in a 1955 survey of a Kampala suburb.[74] Moreover, Muwazi remained a prominent member of the ruling elite and became involved in the controversies that sparked political crisis in the postcolonial period.[75]

Muwazi's particularly prominent position within the social and political hierarchy of Buganda alone might account for the destruction of his house in 1949. Yet the fact that mortality rates associated with severe acute malnutrition remained extremely high in this period meant that Muwazi's work with Trowell involved extracting blood from severely malnourished children who had very slim chances of recovery. Muwazi's professional knowledge and skill, therefore, entailed working closely with British Protectorate officials in a capacity that increased his personal wealth and prestige, without any clear benefits for the children brought to the hospital for treatment and care. Far from suggesting an "unsophisticated" misunderstanding of biomedical procedures, as some argued, connecting blood-taking accusations leveled at Muwazi to his nutritional research reveals a local dissatisfaction with and distrust of the biomedical work that was being carried out on Mulago Hill.[76] Those who targeted Muwazi and destroyed his house sought to critique a depraved form of political leadership in Buganda and the lengths to which some Ganda were willing to go in order to achieve and maintain an elite status. They sought to critique ongoing experimentation on severely malnourished and dying children. Understanding the attack on Muwazi's property as a local indictment of his nutritional work makes it possible to then examine the resulting consequences for the future of nutritional research, treatment, and prevention in Uganda.

"No Survey without Service"

Expatriate physicians and colonial officials routinely disregarded African objections to biomedical practices as ignorance and suspicion rather than valid critique and concern. Yet in the aftermath of the insurrection, they exhibited a far more heightened awareness of the explosive potential of blood work. Clear steps were then taken to reduce local resistance to blood taking procedures and to ensure that future nutritional research proceeded with a much greater degree of caution. Even without fully appreciating the connection between the accusations against Muwazi and the insurrection, the colonial administration and physicians working on Mulago Hill responded to this local engagement with medical work by implementing more ethical research protocols.

Immediately after the insurrection, nutritional research entered what one physician generously referred to as a "rather intensely speculative phase."[77] Officials temporarily suspended further nutritional research, and due to concerns that he "experimented on children," Trowell was passed over for promotion.[78] His junior colleague was appointed the new Professor and Chair of Medicine at Makerere, the pediatric department was moved to a different building, and Trowell was transferred to different ward.[79] In an interview with his daughter many years later, Trowell regretfully acknowledged that his research prior to the insurrection involved questionable experiments:

> At first I didn't realize how dangerous they were—taking blood, and doing other things to the liver, liver biopsies, and so on. In the end I thought, we've certainly lost one case, we may have lost two cases, by this investigation. We didn't realize this when I started. So I cooled off, and said, we can't go any further with this. We're not getting much information out of it, and really all this taking of blood, and the rest of it, is upsetting them too much.[80]

Trowell was not alone in his efforts to "move more cautiously," as he put it.[81] The insurrection prompted a deliberate shift in the practices of nutritional research in Uganda. When the MRC began making arrangements to establish an Infant Malnutrition Research Unit on Mulago Hill in 1951, for example, authorities in Uganda insisted the MRC secretary promise the unit would not conduct "school or institutional trials . . . in such a way as to upset susceptibilities."[82] The MRC researcher who had been sent to Uganda prior to the insurrection to extract pancreatic enzymes even considered relocating due to the "difficulties created by the political situation and local feeling about blood sampling." She chose instead to spend several months testing a less invasive method of taking blood, which, she explained, "was an essential preliminary in this country as procedures involving repeated venepuncture would be doomed before they began."[83] Another physician made a clear reference to Trowell's nutritional research, warning that "extreme caution is necessary, as even finger-pricks are the subject of much suspicion and rumor. It is popularly supposed that Europeans take away African blood and sell it. A rumor of this kind can undo the results of years of hard work."[84]

This shift in research protocols was especially evident when Dr. Rex Dean, an established expert in nutritional science, arrived in Uganda in 1951 to continue research on severe acute malnutrition. At the Infantile Malnutrition Research Unit that he established and directed, Dean implemented a policy requiring that all researchers and physicians working at the unit abide by the maxim "No Survey without Service."[85] For the parents of severely malnourished children brought to the unit, "No Survey without Service" meant an assurance that when their children took part in research, they received cutting-edge treatment and care. This practice was also followed at the unit's rural Child Welfare Clinic where children living in the surrounding region were offered the medical care needed for healthy growth and development as part of their inclusion in studies of nutritional health and wellbeing.[86] Crucially, Dean's implementation of "No Survey without Service" was possible in the early 1950s in a way that it had not been prior to the insurrection. Immediately after he arrived in Uganda and observed the appalling mortality rates associated with severe malnutrition, Dean set to work devising an effective treatment. By the early 1950s, he had succeeded in reducing the mortality rates of the condition from between 40 and 60 percent down to between 10 and 20 percent.[87] In cutting the mortality rates associated with severe malnutrition in half, Dean transformed a condition of almost certain death into one that could be reversed with hospital treatment.[88]

With an effective treatment in place, nutritional research in Uganda entered a new phase. More ethical research protocols and treatment that could save the lives of severely malnourished children meant that there was much to distinguish this work from the research conducted prior to the insurrection. There was, however, one component that continued unabated: blood extraction. Examining blood samples remained a fundamental and routine component of nutritional research in Uganda because it served a critical function as a tool of diagnosis. Due to the edema, or accumulation of fluid in the tissues, and the buildup of fat in the liver and under the skin, weight was an inaccurate indicator of the condition's severity.[89] Assessing the severity of the condition was essential to evaluations of whether or not a therapy was working, and significant research was devoted to the development of accurate diagnostic tools. In Uganda, these efforts focused on possible blood tests and, in the interim, serum protein examinations served as the most accurate measure of protein deficiency in young children. Thus blood extraction continued to be the most routine component of nutritional research on Mulago Hill.

In fact, part of what separated blood extraction in the period following the insurrection from the earlier blood work was that it became so routine. Whereas, prior to the 1950s, blood was withdrawn from a heterogeneous mix of patients by a diverse group of doctors and scientists who had multiple motivations for their many investigations, under Dean this research was largely coordinated and confined to the MRC unit. The research conducted in Uganda during the period of diagnostic uncertainty was far more haphazard and exploratory—unexpected findings prompted additional studies and definitive results concluded one line of investigation only to be replaced by another. However, from the 1950s onward, blood tests became *the* routine procedure performed on all severely malnourished patients admitted for treatment and investigation. As all of the reports and publications confirm, "it [was] usual to bleed each child on the day of admission. . . . The bleedings were repeated every 7 days, but some children were bled twice in the first week. The blood was taken from the internal jugular vein."[90]

These routine blood tests were serial examinations, meaning that they were repeatedly performed on the same child throughout the course of treatment, a period usually spanning at least three weeks and often significantly longer. Serial examinations served to monitor progress toward full recovery, and to fulfil the need for control groups. As Dean and his first biochemist explained: "Blood samples were obtained from a neck vein . . . on admission and at approximately weekly intervals afterwards. The times between taking the samples were sometimes varied to coincide with planned changes of diet. . . . The greatest importance was attached to serial examinations on the same child, who thus acted as his own 'control'."[91] Serial diagnostic serum protein examinations, in the absence of viable controls, made blood extraction the central procedure performed on severely malnourished patients at the MRC for more than two decades.

These routine serum protein estimations were not the primary focus of an investigation, but served as a means of monitoring the condition's severity. Again, nitrogen balance studies provided the best example, as they involved routine and extensive blood extraction. The very young children brought to Mulago for treatment were usually in such a severe state of health that collecting specimens at all, let alone for extended periods, proved nearly impossible.[92] In fact, nitrogen balance studies were not successfully incorporated into the MRC's work until the mid-1950s, when the introduction of a "balance bed" originally devised at the MRC unit in the Gambia suddenly made such studies feasible (see fig. 1.1). In two studies that used the balance bed, the primary investigation concerned urine excretion, and yet, as the researchers explained, "the blood of both boys and girls was studied. The boys were placed on balance beds when they were admitted, and received no food . . . until they were bled, at 8 a.m. the next morning. . . . Blood was taken from the internal jugular vein of all the boys and girls at the end of initial fasting, and subsequently at various times during treatment."[93] Nitrogen balance studies were also used to determine the most therapeutically effective combination of ingredients in Dean's effort to develop a therapeutic

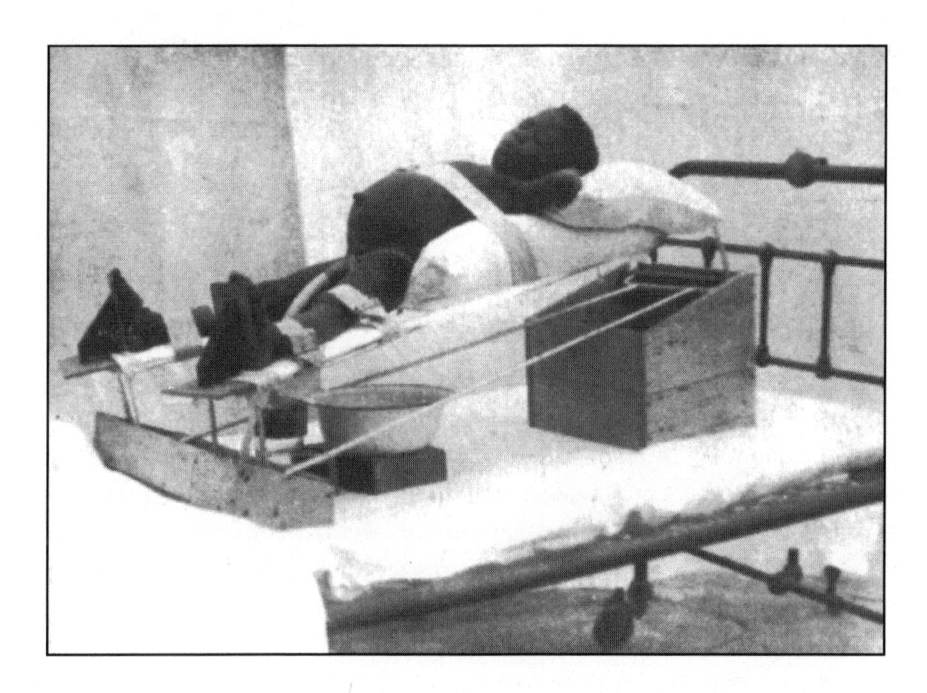

FIGURE 1.1. "Bed for metabolic studies," c. 1952. *Source:* Colonial Office, *Malnutrition in African Mothers, Infants, and Young Children: Report of the Second Inter-African Conference on Nutrition, Fajara, Gambia, 19–27 November, 1952*, 377 (plate 2) (London: H. M. Stationery Office, 1954), by permission of The National Archives.

groundnut (peanut) biscuit, and the discussion of their methodology provides the most detailed description of balance beds in this period:

> The balance beds which have been in use in the [MRC] Unit for several years, allow for the separate collection of urine and feces. . . . A harness around the trunk and legs limits movement but does not entirely prevent it . . . . The accuracy of all balance methods depends to some extent on the regular voiding of feces, which could not, of course, be assured in our children. Extending the length of the periods reduces the importance of inaccuracies, but two four-day periods necessitated a total of fourteen days continuously on the balance bed, and we believed that to be long enough for the children and the staff.

Moreover, as Dean and his colleague noted, "The wards of the Unit have large glass windows, and the children were under continuous observation. . . . Each child was weighed, and bled from the internal jugular vein before and after each period."[94] This continuous extraction of blood as a central feature of nutrition research before and after the insurrection was not without consequences. The advent of a more cautious approach and the development of treatment were crucial if nutritional research on young children was to continue in Uganda, but they could not immediately erase the impact of the questionable experimentation that had been performed on dying children during the period of heightened diagnostic uncertainty in the region.

An Illness of Olumbe

Even with the advent of effective therapy and a more cautious approach, parents of severely malnourished children remained wary of hospital treatment. The damage had been done and local apprehensions did not diminish overnight. People continued to turn to existing remedies and healers first, resorting to hospital treatment often in their final hour of need. This tendency to seek treatment from local healers before consulting a European doctor or biomedically trained physician had been widely observed in this and other parts of Africa since the beginning of colonial rule.[95] Legal sanctions drove local healers underground, but failed to entirely convince people to avoid their services and seek hospital care instead. One physician, who took a special interest in local healing practices, found that even on the eve of political independence, local therapies could be obtained in markets, urban centers, thoroughfares, and near major hospitals and small dispensaries in amounts suggesting extensive and ongoing faith in their efficacy.[96]

This coexistence of local and biomedical forms of healing even led to new categories of illness in Buganda. The word used to designate sickness and disease, *obulwadde*, could be qualified in order to specify whether they were illnesses requiring consultation with local healers (*basawo*) and were thus *endwadde ez'ekiganda*, Ganda diseases, or illnesses that could be treated by a European doctor, known as *endwadde ez'ekizungu* ("European diseases").[97] Not all forms of sickness and disease required treatment, as in the case of the common cold, and not all illnesses could be treated. Forms of debility and disease for which little or nothing could be done were known in Buganda as *olumbe*.[98] The emergence of a category of illness that required biomedical treatment rather than consultation with a healer points to a general willingness to seek hospital therapy when it was proven to work. This was especially evident across East and other regions of Africa with the introduction of highly effective yaws and syphilis treatments. As soon as people saw that a single shot rapidly reversed all visible symptoms, demand for injections skyrocketed. In Kenya such demands exceeded the capacity of existing facilities and treatment camps had to be erected.[99] The popularity of injections for syphilis at the rural maternal and child welfare clinics in Uganda, was, as already noted, substantial enough to generate revenue supporting the work of the CMS-run Mengo Hospital, the largest medical mission station in East Africa.

But not all ailments could be effectively treated in the hospital. As one African medical worker at Mengo was quoted as saying, "My father has worked in the hospital for thirty-five years and he *knows* how many diseases Europeans cannot cope with."[100] Until mid-century, severely malnourished children were either diagnosed as syphilitic and, according to a physician at Mengo, were given "bismuth injections until they would end up in a toxic state with a blue line around the lips," or they were treated with deworming medications or the newly discovered B vitamins, among a range of other largely ineffective forms of treatment and care.[101] Only a small fraction of the severely malnourished children brought to the hospital in the period of diagnostic uncertainty survived. Parents and guardians of malnourished children who turned first to their local remedies were not acting according to an irrational or traditional mind set. Until effective therapies were developed in the early 1950s, they had little reason to have faith in hospital therapy. In fact, prior to the adoption of more ethical and cautious methods, parents and guardians had much to fear. The ongoing centrality of blood work even after the advent of effective therapies and a more cautious approach meant that anxieties surrounding the hospital treatment of severely malnourished children subsided more slowly than might have otherwise been the case.[102]

References to patients "absconding from hospital," "running away," or refusing specific procedures remained frequent through the early 1950s. Often such flight or noncompliance reflected uncertain outcomes, as Dean and others experimented with different therapies. One trial, for example, involved feeding children a variety of locally available foods, and in a number cases the child's condition deteriorated or failed to improve. Parents reportedly and not surprisingly responded by removing their children from hospital care.[103] Another trial, which achieved the highest degree of therapeutic success up to that point, saw over thirteen percent of the children removed from the hospital before making a full recovery.[104] The trepidation with which many parents and guardians approached hospital treatment of severely malnourished children led, at times, to tragic consequences. One child, Mukandekeze, was just two years old when her parents brought her to Mulago Hospital suffering from severe acute malnutrition. Clearly uncertain about the range of procedures performed on malnourished children at Mulago, Mukandekeze's parents refused to allow hospital staff to tube-feed her for very long. After three weeks and with little improvement in her condition, they removed Mukandekeze from the hospital. They continued to take their daughter to a child welfare clinic not far away, but Mukandekeze remained seriously ill and six months later she died.[105]

Not all decisions to remove children from the hospital prior to an official discharge were the result of dissatisfaction with the therapy provided or even unease with specific procedures, although this was often the case. Many parents or guardians of severely malnourished children chose to leave the hospital at the earliest sign of positive therapeutic outcome and their actions may simply reflect satisfaction with treatment and a desire to return home. Parents frequently demanded early discharge as soon as their child's edema dissipated and their appetite improved.[106] One child, Namadu, who had been brought to Mulago for treatment on a number of occasions, was admitted with severe acute malnutrition again in 1952. As soon as Namadu's edema diminished and he showed clear signs of recovery, his parents removed him from the hospital, or in the typical biomedical shorthand of the time, they reportedly "ran away."[107] The vast majority of those who removed their children prior to an official discharge, however, reveal a lingering set of misgivings over procedures performed on severely malnourished children in Uganda. In the examination of pancreatic enzymes discussed above, for example, 20 percent of the children died and as many "ran away" before physicians could extract digestive enzymes a second time.[108]

The decision to bring a severely malnourished child to the hospital for treatment was not a decision parents and guardians took lightly. In addition to transportation expenses, time spent with a sick child in the hospital meant neglecting work and household duties, including the cultivation of food and cash crops, the care of other children, and the collection of water and firewood.[109] Given the burdens of lengthy periods of treatment, parents undoubtedly demanded discharge or simply removed their children from the hospital as soon as recovery appeared certain due to such practical considerations. Yet, a decade after the development of effective therapies, physicians and scientists working with severely malnourished children at Mulago no longer reported that parents refused specific procedures like tube-feeding or removed their children before they were officially discharged. This absence alone is telling. As we will see in the next chapter, those working to treat and prevent severe acute malnutrition in later decades faced a different set of concerns due to growing *demands* for hospital therapy. This contrast reveals that parents of severely malnourished children remained concerned about the questionable research practices for a number of years after they were replaced by a more cautious approach.

Physicians and scientists working in Uganda in this period were fully aware that people lacked confidence in the hospital treatment of severe malnutrition. In addition to their frequent references to patients "absconding from hospital" or running away, they openly acknowledged that malnourished children were rarely brought to them for treatment. According to Trowell and his colleagues, "it is only in exceptional circumstances . . . that children are brought to any hospital because they are suffering from kwashiorkor. . . . They are brought to hospital largely because they have acquired some well-recognized infection." As a result, "expeditions into the villages were necessary to convince mothers that their children . . . had an illness which could be treated in hospital."[110] This reticence to seek hospital treatment for malnutrition meant that throughout this period, children were only brought to the hospital as a last resort and only after a range of local remedies had been tried.

Physicians treating severely malnourished children often found that parents first sought treatment for a number of locally recognized illnesses. The principal one was *obwosi*, a condition signaled not by a specific set of symptoms, but by signs of illness in a child whose mother had become pregnant.[111] The "heat" from the subsequent pregnancy was seen as the cause of illness and in order to prevent or alleviate obwosi, a newly pregnant mother ceased breastfeeding and physically distanced herself from her child by no longer sleeping in the same bed or carrying her child in a sling or *ngozi*.[112] Pregnancy and fears of obwosi were also a pretext for sending a young child to live with an aunt or grandmother.[113] Conditions associated with specific symptoms of severe acute malnutrition included *omusana*, which attributed the lightening skin hue and loss of hair pigment to sun exposure; *obusulo* and *empewo*, which were linked to swelling; and *ekigalanga*, a condition characterized by fever, diarrhea, abdominal pain, appetite loss, and cold feet.[114] Ekigalanga and empewo were both conditions connected to spiritual forces requiring spiritual remediation. Obusulo was an illness caused by seeds entering a child's body and treatment focused on their removal. Children diagnosed in the hospital as severely malnourished often had many small incisions in their skin, at times with a paste containing ash from burnt plantains applied to the cuts or rubbed over their bodies.[115] One such child was observed in the mid-1950s, "encrusted with a grey coating of ashes; her mother was desperate with anxiety for her and was simultaneously arranging to take her to the hospital."[116]

In light of the reasonable fears surrounding hospital treatment of severely malnourished children in this period, many parents and guardians only brought their children to the hospital when it appeared that there was little or nothing that could be done, when it appeared that they suffered from an illness of olumbe. The problem was that by the time severely malnourished children were finally brought to the hospital, they were in such an acute and severe state that they required immediate emergency measures to save their lives. The years of diagnostic uncertainty had taken their toll. Parents were justifiably wary of the procedures performed on severely malnourished children and continued to try existing forms of treatment first. These children often arrived at the hospital desperately ill and so physicians devised the emergency measures needed to save their lives. The diagnostic uncertainty of the early years of nutritional work in Uganda influenced local interactions with hospital treatment in ways that then shaped the form that treatment took. Children were often not brought to the hospital until their condition was a medical emergency. Physicians and scientists responded, as we will see, by medicalizing malnutrition.

In 1949 nutritional research in Uganda was swept up in a political insurrection, and the attack on Eria Muwazi brought this research to an abrupt halt. The insurrection convinced colonial administrators that, in order to avoid further unrest, future nutritional work had to proceed with greater caution. The accusations that Muwazi "kill[ed] children by taking blood" successfully altered the course of nutritional research in Uganda, prompting researchers to devise more ethical procedures, even as they dismissed the rumors of blood taking as unsophisticated and ignorant fears of Western medicine. The fact that blood tests remained a central feature of the research that resumed in the postinsurrection period, without further incident, suggests that blood extraction was not the crucial issue prompting concern. Instead, the accusations leveled at Muwazi were about the ethics of performing dangerous procedures on children who faced almost certain death. Targeting Muwazi was a local indictment of biomedical work that failed to improve health and wellbeing, work that appeared to improve Muwazi's status and prestige at the expense of the people in his care. When colonial officials insisted on more cautious research protocols, they were responding, albeit unwittingly and from a state of nervousness, to the demands of the Ugandan people and their engagement with biomedical work.

Connecting the so-called rumors of Muwazi's blood taking to his medical work on Mulago Hill reveals that local concerns regarding the ethics of this work compelled future researchers to devise new policies, like "No Survey without Service." Any other analysis risks attributing the adoption of these more ethical protocols solely to Rex Dean and his expatriate colleagues, a move that further obscures African agency in a narrative that leaves biomedical ethics an import of the West. It also illustrates how notions of "unsophisticated" fears and "native ignorance" of "Western" medicine miss important local appraisals and critiques of questionable ethical practices.[117] These local appraisals also indicate that people in Uganda had very little faith in biomedicine when faced with severe acute malnutrition, and with good reason. The diagnostic uncertainty that characterized the early decades of nutritional work in Uganda meant that children brought to the hospital suffering from severe acute malnutrition were often subjected to a number of experimental and extractive procedures, even though little could be done to save their lives and the vast majority did not survive. Under these circumstances, parents in Uganda had little reason to bring their malnourished children to Mulago for treatment that did not yet exist and appear to have only done so as a last resort, when nearly all hope was lost. Severely malnourished children were, as a result, brought to the hospital when their illness had become acute, when it became an illness of olumbe. The highly extractive and dangerous procedures that characterized nutritional research in Uganda until the mid-twentieth century allow us to, therefore, see anxieties surrounding biomedicine in a new light. Parents who sought alternative treatments first were clearly not acting according to an irrational and traditional mind set, as has often been assumed. Instead their fears appear now, in retrospect, to be warranted. This early chapter in the history of nutrition and colonial medical research serves as a reminder to both historians and global health practitioners that local responses to medical interventions cannot be reduced to cultural frameworks alone; rather, they must be seen as complex and dynamic historical engagements or "accumulated reflections."

Children brought to the hospital in such a severely malnourished state required emergency measures to save their lives. As will be explored in the next chapter, this local response to nutritional work thereby shaped the measures that the physicians and scientists at Mulago developed in response to the condition, with repercussions for years to come. Thus the diagnostic uncertainty of the early years of nutritional work in Uganda influenced when parents brought their children for hospital treatment, and this in turn shaped the development of that treatment. The history of colonial medicine in this part of Africa represented not a single encounter, but a set of interactions. The shifting local response to nutritional research suggests that people in colonial Africa were not averse to biomedical procedures and care, provided they in fact improved health and wellbeing. This insight is not only essential to an appreciation of the history of colonial medicine in Africa and other parts of the world, but is also important to contemporary health programming. It suggests that particularly when it comes to global health, it is crucial to recall that local engagement with biomedical work is shaped in large part by the residue of past experiences. People engage with health systems in ways that are shaped by long histories of medical research and provision. Evolving practices are influenced not only by the latest scientific developments, but also by the therapeutic decisions of patients and their communities, and their responses to the quality of the care they have received and continue to receive.[118]

2 MEDICALIZING MALNUTRITION

As the story goes, when Rex Dean, the nutritional scientist, arrived in Uganda he was astonished to find severely malnourished children dying at such alarming rates, and so he immediately set to work devising a life-saving treatment.[1] Dean's therapeutic regimen centered on the provision of a high-protein formula that, in light of the foregoing (and future) controversies over the protein hypothesis, garnered much of the attention. Yet the reason severely malnourished children suffered excessive rates of mortality was only partly due to the presumed absence of a high-protein formula to feed them. High case fatality was also tied to the severe state in which malnourished children were brought to the hospital. Local reticence to seek hospital care for severely malnourished children represented, as we have seen, the residue of a period of ongoing diagnostic uncertainty in Uganda—a period characterized by questionable experimentation on dying children. This period of diagnostic uncertainty meant that the children who were brought to the hospital required emergency medical measures just to save their lives. Dean's efforts to do just that served to *medicalize malnutrition*. This medicalization of malnutrition precipitated an era of unwavering faith in the capacity to contend with the problem of severe acute malnutrition. But life-saving curative measures did not prove to be an effective basis for prevention. Local engagement with and interpretation of hospital treatment, especially as it morphed into prevention, led to unintended consequences that further compromised the nutritional health of young children. This chapter explores these developments and how they were obscured and then forgotten. The unintended consequences of medicalizing malnutrition and the resulting scandal were swept under the rug, making the lessons that might be drawn from an analysis of this mid-twentieth-century effort to prevent severe acute malnutrition unavailable to those involved in future efforts to contend with the problem of severe acute malnutrition around the globe.

The Medical Research Council sent Dean to Uganda as a result of his expertise in the prevention of malnutrition. His earlier success developing mixtures of plant proteins "rivaling milk in nutritive value" for malnourished orphans and schoolchildren in postwar Germany appeared directly applicable to the problem of severe childhood malnutrition in Uganda.[2] Dean never lost sight of his goal to prevent malnutrition, but in the end his major contribution to applied nutritional science was the development of a highly effective and highly curative therapy. Moreover, given his mandate to develop a vegetable-based preventive mixture, as he had done in Germany, Dean's high-protein therapy ironically ended up being a milk-based formula that mixtures of vegetable proteins could never rival.

Faced with the startling mortality rates of malnourished children at Mulago Hospital, Dean could not afford to squander time developing a plant-based therapeutic mixture. Once mortality rates fell, he would turn his attention to local sources of vegetable proteins that could become the basis of effective prevention. In the meantime, Dean sought to treat malnourished children with milk simply because dried skimmed milk was the most inexpensive and accessible source of protein in Uganda at the time. As a waste product in the manufacture of butter in Europe and the United States, ample supplies of dried skimmed milk were easily acquired in the postwar period.[3] But skim milk was not without its shortcomings. Although it was not known at the time, many severely malnourished children in Uganda were lactose intolerant and developed diarrhea in response to the skim milk–based formula. Diarrhea is a very common symptom of lactose intolerance, but is extremely dangerous in already acutely malnourished children. One twelve-month-old child undergoing treatment in this early period of therapeutic experimentation developed such loose stools that her weight loss forced Dean to stop her treatment altogether. Fortunately, she did eventually make a full recovery, but her experience and similar reactions among other severely malnourished children indicated that, on its own, skim milk was not a satisfactory form of treatment. Dean dealt with this dilemma by reducing the amount of skim milk and supplementing the mixture with Casilin, a commercially produced preparation of calcium caseinate containing an 80 percent concentration of milk protein. Despite the added cost, this high-protein therapeutic formula was a resounding success. Even before cottonseed oil was added to the formula in order to compensate for the diminished caloric content, Dean and his team in Uganda were able to celebrate the development of the first effective therapy for severe childhood malnutrition (see fig 2.1).[4]

But the development of Dean's high-protein formula was only part of the story. Given that the severely malnourished children brought to the hospital were already in such an acute state upon arrival and had considerably diminished capacities to digest and absorb even essential nutrients, Dean insisted on the institution of what he called "dietary discipline." Coining the term "dietary discipline" emphasized that the provision of dietary therapy in severely malnourished patients was comparable to the provision of drug therapy to treat infection.[5] In the regimented system of infant feeding that Dean developed, a precise amount of protein and calories, determined by the child's weight, was prescribed and administered at specific intervals throughout the day and night. The high-protein therapy was prepared in a glass bottle that in order to avoid spoilage had to be replaced on six-hour rotations.[6] In fact, under Dean's direction all aspects of treatment then became standardized. Secondary infections were so prevalent that, in the

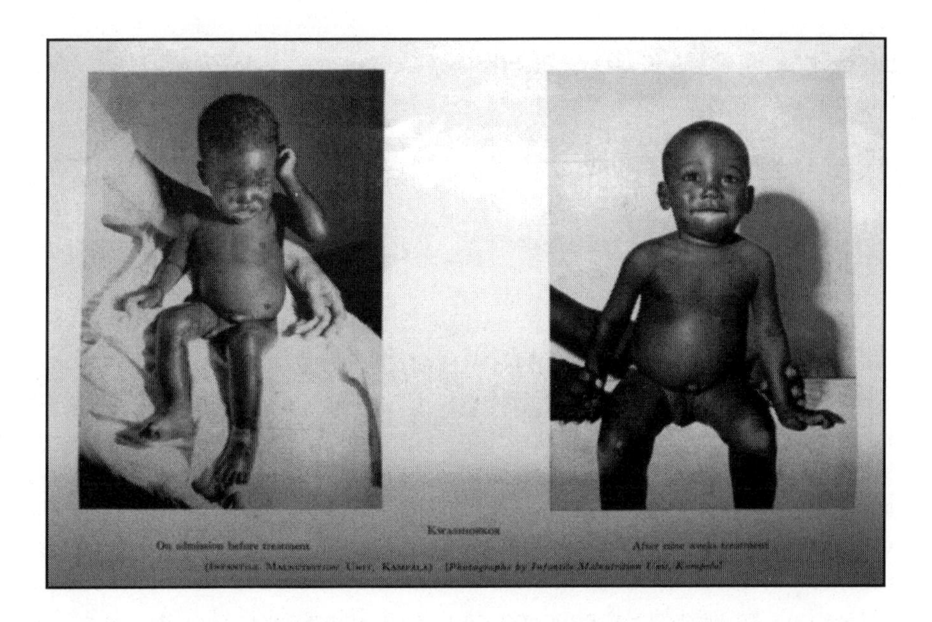

FIGURE 2.1. Child treated for kwashiorkor at the MRC Infantile Malnutrition Unit, Mulago Hill. *Source*: Annual Report of the Medical Department, for the year ended December 31st 1955, Ministry of Health, by permission of the Ugandan National Archives.

initial week of treatment, routine therapeutic measures included daily injections of penicillin, whether or not an infection was evident. Children also automatically received treatment for malaria, anemia, dehydration, and potassium loss.[7] In responding to the severe condition of the malnourished children brought to the hospital, dietary discipline transformed the treatment of severe acute malnutrition into a highly curative, hospital-centered experience involving tubes, formulas, syringes, IVs, and injections (see fig 2.2).

Only two years after arriving in Uganda, Dean could report in the *Lancet* that the concentrated milk-protein formula had already succeeded in reducing the mortality rate to between 10 and 20 percent, a significant achievement given the 75 to 90 percent mortality reported in the 1930s and 1940s by Trowell and others.[8] Biochemical measures of recovery and rehabilitation provided equally compelling evidence of the formula's therapeutic efficacy. Total levels of protein found in the blood, for instance, doubled within one week and reached expected levels for healthy children around the third week of treatment.[9] For a child to achieve a full recovery required the resumption of weight gain and growth at rates that would facilitate the catch-up needed for a stunted child to reach the weight and height considered standard or normal for her age. Only in exceptional cases was it possible to keep a child in the hospital long enough to observe this final phase of rehabilitation.[10] The few children who were treated for extended periods with high-protein

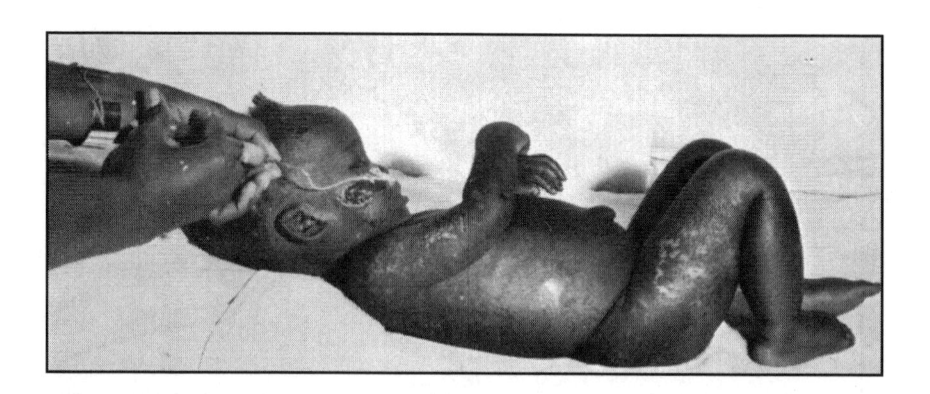

FIGURE 2.2. "Kwashiorkor in a 17 month old Ganda boy, showing syringe feeding . . . through a fine polythene tube." *Source*: D. B. Jelliffe and R. F. A. Dean, "Protein-Calorie Malnutrition in Early Childhood (Practical Notes)," *Journal of Tropical Pediatrics*, December 1959, 96–106, by permission of Oxford University Press.

therapies gained weight at accelerated rates, which over time could eventually reverse their underweight and stunted stature. One child, Bandiho, weighed five kilograms below the American standard when she began her therapy, but grew three and a half times more quickly than normal and began to reach the typical weight for her age after a year of hospital treatment.[11]

Despite the lengthy period required to reach healthy measures of growth, the initial phase of recovery involved a highly visible and striking set of transformations in a child's condition, all of which occurred at a phenomenal pace. Even in the very severe cases that were brought to the hospital, nearly all of the most prominent symptoms began to improve within ten days and in some cases by the end of the first week. The anorexia that frequently made intragastric tube feeding necessary subsided so rapidly that children rarely had to be tube-fed the high-protein formula for more than two days.[12] The edema also promptly diminished, as did the rash or dermatosis and the fatty buildup beneath the skin. After only one week of Dean's treatment, children who had been listless and apathetic began to clearly take an interest in their surroundings, and this improvement in their demeanor was interpreted as a clear sign that they were on the road to recovery.[13] The formula's capacity to rapidly and visibly resuscitate children who had been very near death did eventually contribute to shifts in local perceptions of hospital treatment even if the ongoing blood work at Dean's MRC Unit on Mulago Hill meant that such shifts took longer than might otherwise have been the case.

In time local concerns over experimental procedures and reports of parents refusing treatment and absconding from the hospital gave way to signs of increasing acceptance of hospital therapy. By the early 1960s, if not before, growing local confidence became outright demand. Thus the development of the highly effective and highly curative emergency measures capable of saving the lives of severely malnourished children ushered in a distinct turning point in local views of and engagement with biomedical treatment of severe malnutrition. One of the physicians working in Uganda in this period, Mike Church, wrote for example that "the dramatic intravenous and intragastric therapies, which were often lifesaving, were expected by mothers. In fact, the fame of the hospital resulted in some mothers traveling hundreds of miles" in order to obtain treatment for their severely malnourished children.[14] What had been an illness of olumbe, a condition for which there was no hope, became something else. What had been an illness prompting physicians and scientists to perform a myriad of highly extractive and experimental procedures on children who nonetheless died became a condition for which routine emergency measures could all but guarantee recovery and survival. No longer did physicians write of patients absconding from the hospital. Instead, the medicalization of malnutrition, the effective response to the severe condition in which children arrived, led to a growing local demand for life-saving hospital procedures.

The development of a novel therapy also signaled a new disease entity in both biomedical and Ganda diagnostic registers. For the biomedical community, the success of Dean's high-protein therapy appeared to prove that the condition was a form of severe protein deficiency and, recognizing Williams's earlier insight, the Ghanaian term kwashiorkor became the internationally recognized name for the condition. The condition for which Ugandan parents traveled great distances to obtain hospital treatment was known, for a brief period, as *olbuwadde bw'eccupa* or "bottle disease." Like its Swahili counterpart, *eccupa* (pronounced "chupa") is the Luganda term for "bottle" and the notion of eccupa disease reflects the bottles and tubes that were so central to the treatment of severely malnourished children. As Church later remembered,

> We actually discovered that they had created a new mythology, they had a new word for kwashiorkor . . . they called it eccupa disease. Eccupa being the bottle and of course the bottle was the intravenous and intragastric feeding. So when they went into the pediatric wards they would immediately be put on drips, which would be intragastric feeding and intravenous fluids, and the mothers watched this with great wonder because of course, in the wards, that was what transformed them.[15]

In Ganda diagnostics and etiology, this transformation confirmed the diagnosis: olbuwadde bw'eccupa was a condition that required hospital treatment and a veritable barrage of therapeutic measures centered around bottles containing prescribed amounts of Dean's skim milk formula. A new and effective treatment indicated the presence of a new disease, one that for both biomedical practitioners and Ganda observers was seen to require extensive and immediate medical attention. Whether known as kwashiorkor or eccupa disease, this medicalization of malnutrition was to have far-reaching consequences for child health and welfare, especially as it shaped both international perceptions of the condition, the resulting programs of prevention, and local engagement with these preventive measures.

Even before Dean's high-protein formula appeared to provide the final confirmation that severe acute malnutrition was caused by protein deficiency, the condition was seen as a worldwide scourge demanding intervention. The international attention on protein malnutrition during the 1950s was such that one expert claimed that "in human nutritional studies and in international public health this has been a protein decade."[16] This view was also echoed by the head of the nutrition department in Bombay, who wrote, "We have moved from the era of vitamin research to the era of protein research."[17] The international response reflected particular interpretations of the mounting evidence implicating protein, and much of that evidence emerged from Uganda. The medicalization of malnutrition on Mulago Hill not only launched the "protein decade," but continued to have an influence for many years. When the Joint FAO/WHO Expert Committee on Nutrition held its second meeting in 1952, the proceedings were dedicated entirely to the condition and Trowell, Davies, and Dean presented the findings of their latest research.[18] Suddenly thrust onto the world stage as an international center of nutrition research, Dean's MRC Infantile Malnutrition Research Unit was in a position to shape how the condition was understood and what was to be done about it. Moreover, the fact that the protein decade coincided with the postwar development era was far from coincidental. Efforts to contend with the problem of protein malnutrition reflect the international faith that was placed in scientific solutions to the problems facing so-called developing world regions. The potential promise of Dean's high-protein therapy, its simplicity and visibly transformative impact on child health, emboldened those persuaded by the proverbial magic-bullet, one-size-fits-all approach. It was in this way that a specific framing of the problem of protein malnutrition, temporarily at least, foreclosed alternative ways of promoting nutritional health.

The first move in the increasingly narrow and highly medicalized definition of the condition was to confine the problem of severe malnutrition solely to young children. Initially people of all ages were included in studies of severe malnutrition and the steady stream of immigrants who came from present-day Rwanda and Burundi and arrived in severely malnourished states were, as we have seen, an important part of early studies of nutritional health in Uganda. In fact, research on protein malnutrition in adults was so central to the work carried out in Uganda that an entire part of the seminal text that Trowell, Davies, and Dean published on kwashiorkor in 1954 was devoted to protein malnutrition in adults and the symptoms observed in adult cases were not regarded as entirely distinct from the infantile syndrome.[19] With the advent of an increasingly medicalized vision of kwashiorkor as a medical emergency, the focus shifted to young children. It was the WHO's seminal report, *Kwashiorkor in Africa*, that first narrowly defined the condition exclusively as a childhood illness. Even while acknowledging that "a syndrome very similar to kwashiorkor is undoubtedly encountered in other age-groups and even in adults," the authors of the report argued that the condition known as kwashiorkor should be confined to children and especially to children in the weaning phase of life.[20] The rationale was that the protein requirements for growth and development were higher in children under five years of age than at any other point in the life cycle.[21] Young children were thereby particularly susceptible to severe protein deficiency and the condition posed a much greater threat to their survival. This pronounced prevalence, severity, and mortality made young children an obvious and understandable focus of medical research and attention.[22] With the publication of the 1952 WHO report, the medicalization of malnutrition, in which children were not brought to the hospital until their condition had become so severe that they required emergency measures to save their lives, came to therefore define the condition and circumscribe the resulting international response. Rather than a broad public health concern requiring comprehensive interventions, this narrow definition of the problem prompted a far more limited and targeted solution.

The exclusive focus on young children privileged explanations emphasizing how and what children were fed. Evidence that the condition was exceptional in infants under six months old was attributed to breastfeeding, and it was frequently observed that exclusively breastfed infants in Buganda grew at rates that met and even surpassed US standards.[23] All of that changed, however, when children reached the age of six months and breast milk no longer sufficed as their sole source of nourishment. The subsequent decline in the nutritional health of Ganda children was partly attributed to the culture of weaning in Buganda. Customs like the practice of sending newly weaned children to live with a member of the father's clan were frequently cited.[24] Mothers were also specifically criticized for arbitrarily choosing the date to cease breastfeeding and for the subsequent change in their attitude toward their children. A psychologist working at the MRC Unit argued, for instance, that "the mother does not only stop giving him the breast, but often behaves as though she is deliberately trying to effect a separation";[25] "she no longer carries him on her back, or sleeps with him, no longer consoles him, but laughs at him if he cries or merely tells him to stop crying."[26] The so-called "maternal deprivation" that a newly weaned child supposedly felt as a result of these "abrupt" weaning practices was thought to contribute to the anorexia of malnourished children—a refusal to eat might, it was hypothesized, be an attempt to express the distress of being suddenly separated from one's mother. Describing the weaning practices in Buganda as a "psychological trauma" precipitating severe protein malnutrition, physicians and scientists blamed Ugandan women and portrayed them as an essential part of the problem.[27] For a time such attitudes, thereby, limited how mothers figured in public health campaigns designed to prevent severe malnutrition.

Ugandan mothers were also faulted for not ensuring that their young children consumed sufficient quantities of protein. Ignorance of the unique nutritional needs of young children was frequently noted, as well as cultural taboos, which, in proscribing certain foods for women, did not officially apply to children, but were thought in practice to mean that most of the meat, fish, and eggs were reserved for men.[28] The MRC's nutritionist was especially critical of the diet consumed by children once they had been weaned. "It is the custom," she wrote, "for young children after weaning to have the same foods as the adults in the family, the idea of preparing food specially for the children being completely alien to the Baganda."[29] The adult diet was deemed inadequate primarily because the high-protein relishes and sauces, or *enva*, were not thought to be offered in a manner guaranteeing that children actually consumed them. According to this view, children in Buganda were neither actively fed nor considered physically capable of consuming adult food in the customary fashion. Protein deficiency was, therefore, understood to be partly due to the way in which food was consumed, using the hands to dip an edible portion of the primary staple into the protein-rich sauce or relish.[30] Physicians and scientists also criticized the manner in which a meal was served, finding fault in the infrequency of meals—a factor curiously attributed to cultural practices rather than economic constraints—and the customary practice of sharing food from communal dishes. Although it was noted that individual portions were becoming increasingly common, especially for the father or household head, the fact that women and children continued to eat from communal plates was viewed as another factor limiting the amount of protein consumed by the youngest members of the family.[31]

The fundamental problem with dietary practices in Buganda was that children were reportedly weaned onto foods that contained very little protein. The predominant staple of Ganda cuisine, which bore the brunt of this critique, was the green cooking banana or plantain, known as *matooke*, followed in importance by sweet potatoes, or *lumonde*, and cassava. Indictments of these principal components of the local diet unduly focused on the translation of Ganda conceptions of food, thereby overlooking the importance of the accompanying sauce or enva. As in many African societies, the definition of food in Buganda was synonymous with the primary staple; matooke best fulfilled the purposes of eating, which were to satisfy hunger and fill the stomach. Yet no meal was complete without enva, and these sauces or relishes were usually composed of vegetables as well as groundnuts, beans, or, when available, meat or fish, and could therefore be very high in dietary protein. Overlooking the importance of these protein-rich parts of the customary meal, biomedical personnel focused on what they saw as the troubling cultural importance of matooke in Buganda. The very low ratio of protein to calories found in Ganda staples meant that, as biomedical practitioners saw it, the amount of food containing sufficient nutrients was far too large for the small stomach of a very young child.[32] The MRC Unit's nutritionist claimed, for example, that "everything about '*matoke*' is ritual of an intensity that is almost religious,"[33] and in his first treatise on the condition after his arrival in Uganda, Dean concluded that the problem of protein malnutrition could easily be attributed to a simple "fondness for matoke."[34] As a result, early efforts to inform Ugandan mothers of the nutritional needs of young children focused on convincing them that matooke was a bad food for young children. Given that nearly all children in Buganda grew up eating matooke and not all developed severe acute malnutrition, this not only was an unconvincing message, but represented a missed opportunity to highlight the protein-rich components of the sauces served as an essential part of every meal.

On the international stage, this framing of the problem of protein malnutrition influenced the actions that were taken. Women, as mothers, were either blamed for flawed weaning and feeding practices or they were portrayed as victims who together with young children were denied protein-rich foods. The emphasis on protein malnutrition therefore drew much-needed attention to maternal and child health and the nutritional requirements of pregnant and nursing women and their young children, but did so in a manner foreclosing the possibility of enlisting and empowering women as a part of the solution. What is more, the particular emphasis placed on the protein-poor quality of dietary staples framed severe acute malnutrition as an inherent absence, as an inadequacy, and a gap. Entire world regions were categorized as fundamentally lacking in dietary protein, and the prevalence of severe acute malnutrition became evidence of a "worldwide protein gap." To address the dearth of available protein in parts of the world, in 1955 the WHO created a Protein Advisory Group, and in cooperation with the Committee on Protein Malnutrition of the US National Academy of Sciences/National Research Council, the Protein Advisory Group monitored global protein supplies and led efforts to close the "protein gap."[35]

The principal determinant of regional protein supplies was seen, first and foremost, as a natural and largely static feature of the environment. Starchy staples associated with tropical areas of the world, namely cassava, plantains, bananas, yams, and sweet potatoes, were compared unfavorably with those of more temperate regions, especially cereals and grains such as wheat. According to one report, these starchy staples supplied a meager 4 percent of global dietary protein while cereal crops accounted for nearly half of global supplies.[36] The inherent inadequacy of tropical staples was further compounded by what was uncritically seen as a natural lack of animal protein in the tropics. Inadequate supplies of cow's milk in tropical environments were depicted as especially critical to the problem of severe malnutrition, as cow's milk was widely viewed, in this period, as an ideal dietary supplement following the cessation of breastfeeding. Evaluating protein supplies as inherent to specific environments defined protein malnutrition as a condition that was endemic in parts of the world. According to the Protein Advisory Group, "A doctor in western Europe might pass his entire career without ever examining a child suffering from one of the many forms of malnutrition so common in tropical and subtropical areas."[37]

Protein deficiency thus became another shortcoming of tropicality that mapped onto notions of the "developing" world as unhealthy and technologically inferior. Moreover, such conceptions fit with mounting evidence from x-rays, nitrogen balance studies, and charts of standard growth rates indicating that a mild form of protein malnutrition preceded the acute stage of the syndrome and that mild protein deficiency appeared to be far more widespread than its more acute cousin. It was thus that what the Protein Advisory Group described as the "recognition of great differences between protein supply and demand in the developing countries"[38] constructed this gap as a fundamental explanation, as opposed to a symptom, of the problems plaguing impoverished world regions. In fact, many began to openly propose that the endemic problem of protein deficiency in Africa and other so-called developing regions perhaps explained their "backwardness."[39]

Alarming reports of insufficient dietary protein across vast global regions remained legion throughout the 1950s and 1960s. The Protein Advisory Group, for instance, warned that "protein supplies in the early 1960s fell short of requirements in most of the developing countries. Predictions for the future . . . continue to be pessimistic for most parts of the developing world." In sub-Saharan Africa the situation was especially dire as estimates indicated that protein supplies fell short of the need by an estimated 270,000 tons.[40] In fact the perception that the "deficiency of protein in the diet was the most serious and widespread nutrition problem in the world" became so pervasive that in 1968 the General Assembly of the United Nations even considered the creation of a United Nations Protein Board—a proposition only dismissed as unnecessary in light of the already existing Protein Advisory Group.[41] In the report the Protein Advisory Group published in 1970, entitled *Lives in Peril*, protein malnutrition was described as "the biggest menace to the future health of the nations in developing regions."[42] Yet, significant efforts had already been made to fill the worldwide protein gap, efforts driven largely by a medicalized vision of severe acute malnutrition and how best to prevent it.

Filling a Gap

Until the mid-1950s, the international campaign to combat protein malnutrition focused primarily on the "emergency action" that the initial report on kwashiorkor in Africa called for—namely a public campaign to spread awareness of the condition and the therapeutic efficacy of the high-protein, skim milk–based formula. As part of this early emphasis on curative measures, it was recommended that "skim-milk powder should be made available to hospitals and maternity and child-welfare centres" and that UNICEF should coordinate with governments to secure supplies of skim-milk for both curative and preventive purposes.[43] The distribution of skim-milk powder as a means of preventing severe protein deficiency was always seen as a temporary measure.[44] By 1955, the agenda shifted from short-term curative initiatives toward prevention. In international reports the vital need for educational programs designed to improve methods of infant feeding figured prominently, but the main thrust was the development of milk substitutes or protein-rich foods specifically formulated for infants and young children.[45]

The protein-rich food program—designed to fill the worldwide protein gap—became a nascent industry centered on the research and development of commercially acceptable high protein mixtures and ingredients. Both the WHO and FAO supplied ongoing finances and consultants to a number of international research centers involved in producing high-protein foods, including Guatemala's Institute of Nutrition for Central America and Panama (INCAP) and the MRC Unit in Uganda. Such funds were augmented by $860,000 in grants from the Rockefeller Foundation and UNICEF specifically earmarked for "research in protein-rich foods."[46] The apparent success of these endeavors led the United Nations to eventually call for a $75 million program.[47] Not surprisingly, Dean was at the forefront of this work from the very beginning, and, with assistance from the FAO and the WHO, the MRC Unit on Mulago Hill coordinated with INCAP to lead the development and commercial production of mixtures suitable for supplementary infant feeding.[48]

Dean spent the early 1950s conducting promising trials with mixtures containing soy milk and sweet banana. In 1952, he wrote to Cicely Williams to announce "that we have just discharged from hospital the first child who has been successfully treated for fairly advanced Kwashiorkor by means of an all-plant diet. . . . We are pleased with this result because we think it proves that we are on the right lines: and that prevention cannot be impossibly difficult. There is plenty of soya here."[49] Further trials revealed that although the original high-protein formula continued to be the most effective therapy in particularly acute cases, children who received the soy preparation recovered as well as those treated with milk proteins.[50] Enthusiasm over the preventive promise of soy quickly dissipated, however, when Dean and his colleagues discovered that soy beans were not grown for local consumption, but as cash crops in Uganda. According to Ephraim Musoke, who cultivated soy for a brief period following World War II, there was little interest in eating soy beans due to the inordinate amount of time and charcoal consumed in their preparation.[51]

In the end, non-milk-based formulas met with very limited success. In the late 1950s, Dean abandoned the aim of devising an entirely milk-free mixture and began experimenting with formulas containing various quantities of milk powder. The primary ingredient of these mixtures was the groundnut or peanut. Unlike soy, groundnuts were grown in nearly every garden plot and had long been an integral component of the local diet. Mixtures of groundnuts, maize and wheat flour, cottonseed oil, sugar, and skim-milk powder proved to be just as effective as the high-protein therapy in hospital treatment—suggesting to Dean and his colleagues that the mixture could also reduce the overall prevalence of protein malnutrition. By making the ingredients into a cooked biscuit that was later ground into a fine powder, it was thought that the meal could be stored in even "unsealed containers without apparent deterioration for many months."[52] As Dean noted, "a dry material has great advantages in an equatorial climate."[53] In 1959, plans were put in motion to produce the groundnut biscuit commercially through a local Ugandan company under the guidance of the biochemist at the MRC Unit and an expert sent by the WHO.[54]

Already by 1962, the Expert Committee on Nutrition could claim, "the stage has now been reached when it can be said with some confidence that one of the major, most promising and most rapid methods of improving nutrition in many countries lies in an effective and speedy programme for the production, distribution and consumption of protein-rich foods."[55] Confidence, in the early 1960s, appeared warranted. Commercial production of flours made from soybeans, groundnuts, cottonseeds, and refined fish had already achieved satisfactory results and acceptability trials of these products and mixtures made from them were underway in more than twenty different countries across the globe.[56] By the later 1960s, a number of different high-protein mixtures had entered the market as distinct products; from INCAP, a product known as Incaparina, Fortifex in Brazil, Pronutro in South Africa, Arlac in Nigeria, Lac-Tone in India, and Surpro in East Africa. And at the end of 1968, 162 tons of Superamine were ready for the market in Algeria.[57]

Yet these high-protein mixtures largely failed as weaning foods capable of preventing the condition. A series of technical setbacks proved insurmountable. In Uganda, all work on the groundnut biscuit was immediately suspended when it was discovered that a highly toxic and carcinogenic mold, aflatoxin, was frequently found on cereals and oilseeds, including groundnuts.[58] In other regions, high production costs made high-protein foods prohibitively expensive for the poorer segments of the population most susceptible to protein malnutrition. Early success with Pronutro in South Africa, where approximately 600 tons were sold per month, diminished when the rising cost of ingredients was passed onto consumers who could no longer afford the product.[59] A similar fate befell Incaparina. After nearly two decades of production in some countries, the director of INCAP insisted that without "subsidized distribution" or significant poverty reduction, "a commercially produced and marketed vegetable mixture" could not be expected to "'solve' the problem of protein-calorie malnutrition"—an insight worth emphasizing in light of the early twenty-first-century enthusiasm for very similar products (as exed in the epilogue).[60] Prohibitive costs were compounded by the failure of acceptability trials or poorly planned initiatives such as the fortification of bread in South Africa with over 800 tons of fish flour only to discover that brown bread was not a significant component of the diet among poor African children.[61]

When the discovery of aflatoxin left milk as the only source of protein considered safe and effective, Dean focused his energies on expanding the distribution of dried skim milk through the commercial production of "reinforced milk packets." Dean's decision to abandon his efforts to develop a plant-based mixture after the failure of the groundnut biscuit was also influenced by his own deteriorating condition. Dean contracted transverse myelitis in 1956 while conducting a survey on kwashiorkor in Malaya for the WHO, and in the early 1960s the illness began to take its toll.[62] The local commercial production of fortified and packaged dried skim milk was Dean's final effort to contribute to the prevention of kwashiorkor in Uganda. Equivalent to one pint of full cream milk, "reinforced milk packets" contained a simple mixture of dried skimmed milk, cottonseed oil, and cane sugar. After a series of trials indicated positive results in treatment and prevention, Dean began to work with a local company, the "'Uganda Co-operative Creameries' in order to produce this 'reinforced milk' on a commercial basis."[63] Prior to his death in 1964, the central Ugandan hospital reported that dried skimmed milk was "either issued plain in bags . . . or reinforced with additional calories . . . 'reinforced milk packets' as they are called."[64]

Parents who received dried skim milk at Mulago were reportedly given explicit instructions to add it to a child's meal and not to reconstitute the powder as milk for the child to drink. A label created to accompany the reinforced milk packets instructed, in fine print: "Suckle your child until he is one and a half years old. In addition, from the age of six months onwards, give him some solid food such as eggs, beans, peas, groundnuts or green vegetables, also if possible some meat or fish."[65] However, the message conveyed in bold print on the label read "EMMERE Y'OMWANA EMMERE EKUZA OMWANA OBULUNGI" or "Food which makes the child grow well"[66] (see fig 2.3). The policy followed at the outpatient department was to "supply the child's mother with enough of these packets to last her until she next comes up, and, rather than ask her to dissolve the milk powder in water of doubtful safety, ask her to add it to the sauce that she would normally give her sick child, or to other items of his diet."[67]

Thus dried skimmed milk was either given or sold at hospitals and health centers in Uganda beginning in the early 1950s as an immediate, emergency

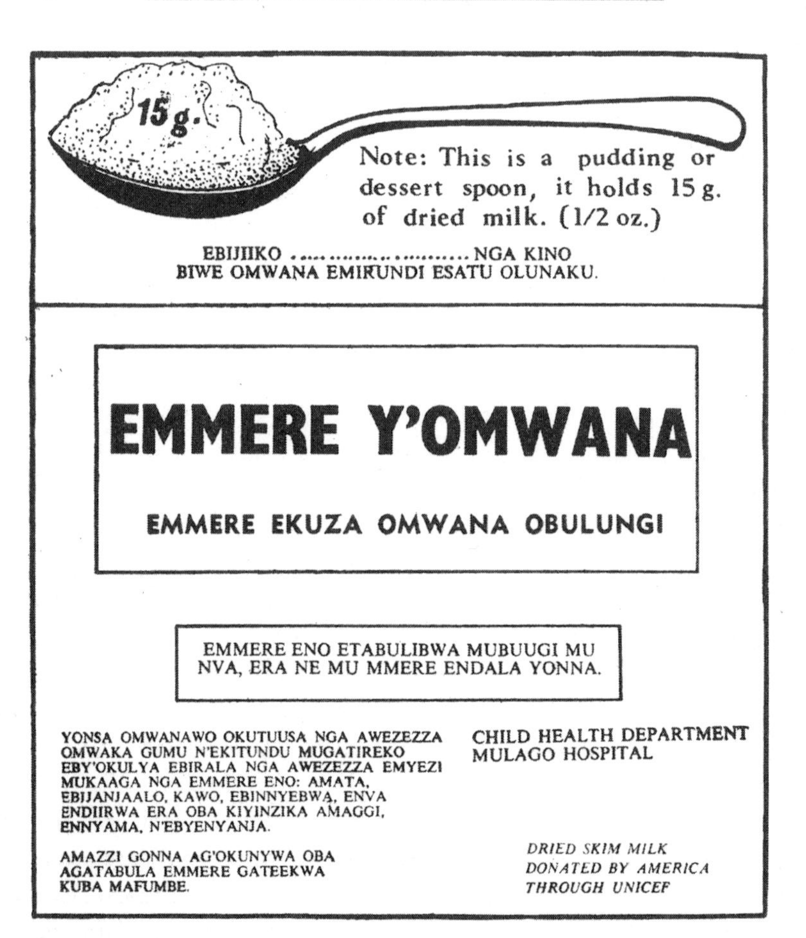

FIGURE 2.3. Label for milk packets. *Source:* "Protein Calorie Malnutrition," in *Medical Care in Developing Countries: A Primer on the Medicine of Poverty and a Symposium* from Makerere, ed. Maurice King (Nairobi, Ken.: Oxford University Press, 1966), 14:14.

measure until high-protein foods could be developed. The failure of the high-protein food program meant that the practice continued in Uganda through the late 1960s when physicians at Mulago decided to begin "weaning the mother from free milk powder."[68] In the interim, Dean and the prominent professor of pediatrics, Derrick B. Jelliffe, recommended that in less severe cases, "Dried skimmed milk (D.S.M.) should be supplied to the mother either once or twice weekly, or fortnightly" unless the distribution of smaller quantities proved impractical due to distance from the clinic, in which case they advised the "monthly issue of powdered milk to mothers."[69] To prevent malnutrition in children not yet presenting symptoms, they suggested "that a supply of special infant food, such as D.S.M., should be made available at child welfare centres for the majority of poorer parents."[70] As the Uganda Government Nutrition Advisor, Dean was in a position to advocate that the policy of distributing dried skimmed milk be implemented throughout the country. At mobile child welfare clinics that were held on a regular basis in and around the Ugandan capital, Kampala, Dr. Hebe Welbourn sold dried skim milk to parents who participated in lessons on "mixing dried milk" that were routinely offered.[71] Welbourn advocated "buying the milk in large quantities wholesale and selling it at cost" in order to ensure that the practice of feeding the milk to children became a "routine" part of feeding practices. "If the milk were dispensed [for] free," Welbourn reportedly reasoned, "people would rely upon the gift passively and would cease using it when later it was no longer available as a free service. If, however, they could become accustomed to purchasing it, a demand would be established that might persist."[72] However, free distribution continued at hospitals and health centers, and in 1958 missionary institutions planned to begin weekly distributions of a half-pound of milk powder per child, requesting an initial shipment of 148 tons from the Church World Service.[73]

It was thus that in the 1950s and 1960s parents and guardians who sought treatment for malnourished children in Uganda often returned to their homes with much healthier children and a supply of dried skim milk. Health care professionals provided skim milk powder to forestall the possibility of relapse in children who recovered from malnutrition and to prevent the onset of severe malnutrition in children not yet exhibiting clinical symptoms. With skim milk powder as the basis of both preventive and curative measures, the distinction between prevention and cure broke down. Interviews with elderly women and men in the area around the Luteete Health Center confirm that milk powder was distributed at hospitals and health centers in areas outside the capital city. Two of the women interviewed received milk packets from Mulago and remembered the accompanying label. When asked how they were instructed to feed the skim milk powder to their children, they emphasized the need to boil the water used to make the powder into milk for their children to drink. Not one informant could recall being instructed to add the powdered milk to a child's food.[74] Moreover, the message was essentially "do as I say not as I do," as reconstituting the milk powder into a formula that children could drink was precisely how physicians treated malnourished children in the pediatric wards of Mulago Hospital.

The True Fiasco

The unintended consequence of skim milk distribution in Uganda began to emerge by the mid-1950s as bottle feeding was clearly on the rise. Even supplementary bottle feeding displaced breastfeeding and increased the prevalence of undernutrition or marasmus. Biomedical practitioners were well aware that without access to clean water, bottle feeding posed a significant hazard to the health of young children. Gastroenteritis often followed the introduction of bottle feeding in resource-poor settings and was widely known to be a leading contributor to marasmus or undernutrition. As a result, health education that emphasized the importance of breastfeeding was considered an essential component of preventing malnutrition. Yet health education geared toward the prevention of protein deficiency sent a contradictory message: convincing mothers that breast milk was nutritionally insufficient and required supplementation when a child reached the age of six months engendered significant concern among mothers about the general adequacy of their breast milk. An American psychologist, Mary Ainsworth, conducted a study of infant care in Uganda in the mid-1950s and inadvertently documented how skim milk distribution converged with hospital treatment and health education in this period. Her exceptionally detailed accounts of individual children and the testimony of their mothers and caretakers illustrate the resulting impact on child health and welfare in Uganda. Anxieties about breast milk and the importance of supplementing the diets of very young children became pervasive. In fact, Ainsworth claimed that "No other aspect of infant care was of more concern to the mothers . . . than supplementary feeding, and nearly all of them asked repeatedly for information and advice."[75] Moreover, Ainsworth found that "in every case of dwindling or insufficient milk supply the mother had complained of this before beginning supplementary feedings of milk."[76] This evidence was further corroborated by others. Welbourn, for instance, reported that by the mid-1950s, "most mothers said they started bottle feeding because they had no breast milk."[77] The primary reason, other than pregnancy, that mothers gave for their decision to bottle feed their infants was their lack of confidence in their ability to breast-feed "successfully."[78] Dr. Josephine Namboze, a Ugandan and the first female doctor in East Africa, also found that women in the mid-1960s were concerned about their breast milk. She cited testimony wherein mothers stated that the "breast milk was not enough," or that they were "getting worried [that their] milk did not suit the baby because he was constantly getting sick."[79]

Several of the children included in Ainsworth's study illustrate how fears surrounding the inadequacy of breast milk translated into supplementary bottle feeding with negative consequences for child health and nutrition. Magalita, whom Ainsworth described as thin compared to other children her age, was apparently born in the hospital weighing a healthy 7 pounds and 4 ounces. Magalita also reportedly "thrived" for the first couple months of her life. According to Ainsworth, Magalita's mother thought that her breast milk was inadequate from the outset. "Just how much reality basis there was for the mother's feelings that her milk had always been insufficient," Ainsworth wasn't sure, "but she always felt that this was so." According to Ainsworth,

> She had been a regular visitor to Dr. Welbourn's clinic. Her milk, scanty from the beginning, seemed really to begin to fail. She stopped Magalita's daytime feedings, intending to conserve her milk for nighttime feedings. Dr. Welbourn explained to her that she needed to feed the baby at times throughout the day in order to stimulate the flow of milk, so daytime feedings were reinstituted. Supplementary milk was given as well. At first the mother used a bottle, but she found it too troublesome to sterilize bottles and used a cup instead.

Magalita then reportedly suffered from repeated bouts of diarrhea, and at times fever and a cough. Her weight at forty-three weeks was 15.6 pounds, but over the course of the next four weeks she lost weight and Ainsworth concluded that "the mother's concern about her health and physical development seemed justified."[80] When Ainsworth saw Magalita again just before her first birthday, she was "quite ill." Even after seeking treatment in Kampala, she did not seem to be recovering and she lost additional weight. In addition to a fever and cough, Magalita then developed an earache and reportedly weighed only 15.5 pounds. "A week later," Ainsworth found that "she was still ill—with fever and diarrhea. She was said to be weak, and she looked weak"[81] (see fig 2.4).

Petero, another child observed by Ainsworth in this period, also suffered from poor nutritional health and repeated bouts of illness in the context of anxiety over breast milk and the introduction of bottle feeding. When Ainsworth first met Petero and his mother, he was eighteen weeks old and suffering from a diarrheal disease. At six months, the doctor became concerned that he was not gaining weight and recommended that his mother supplement his diet with solid foods. Ainsworth again visited Petero's family and

FIGURE 2.4. Magalita at fifty-two weeks. *Source:* Mary D. Salter Ainsworth, *Infancy in Uganda: Infant Care and the Growth of Love* (Baltimore: Johns Hopkins Press, 1967), 158 (image 6). © 1967 The Johns Hopkins Press. Reprinted with permission of Johns Hopkins University Press.

found Petero and his siblings sick with fever and diarrhea. According to Ainsworth, "His mother had concluded that her milk was insufficient to nourish him and had supplemented breast feedings with milk from a nursing bottle. He much preferred the breast, and objected to the bottle. . . . At the clinic the following week he was found to have lost a little weight; he weighed 14.4 pounds."[82] Then when Ainsworth went to see Petero again, only a month from his third birthday, she learned that he had been very sick with "fever, diarrhea and vomiting."[83] His mother had taken him to Kampala for treatment, and after his condition worsened and he fainted, she took him to the hospital where he was admitted for treatment. Dissatisfied with his care, the following day Petero's mother removed him from the hospital and took him to a private practitioner where he was reportedly given "injections and glucose." According to Ainsworth, his mother remained very concerned about his condition and although she indicated a week later that he was improving, he still did not appear to be well (see fig 2.5).[84]

Nor were Magalita and Petero the only children included in Ainsworth's study to illustrate how skim milk distribution undermined confidence in breastfeeding and encouraged supplementary bottle feeding in Buganda. A set of twins, Waswa and Nakato, also suffered from "severe and persistent diarrhea" following the introduction of supplementary milk feedings. As their mother's milk supplies subsequently diminished, recurrent bouts of gastroenteritis had a devastating impact on their growth—leaving them in a state of severe undernutrition (see fig 2.6). Another child, Suliamani, exhibited healthy weight gain prior to attending Welbourn's Child Welfare Clinic and beginning supplementary bottle feeding. He then gained less than a pound over the following two months and at thirty-eight weeks "became quite ill with fever, diarrhea, and vomiting."[85] Moreover, all of the elderly informants consulted for this study remembered bottle feeding as a pervasive practice in Uganda during the period coinciding with skim milk distribution.[86]

The difficulty that Magalita, Petero, and the other children experienced gaining weight and maintaining healthy rates of growth in the midst of repeated bouts of illness attest to the known dangers associated with bottle feeding. Moreover, their mothers' testimony indicates that supplementary bottle feeding was a response to fears surrounding the inadequacy of breast milk in the context of repeated consultations with biomedical personnel. Bottle feeding did pose very real threats to child health and welfare, and, as early as the mid-1950s, physicians became increasingly troubled by the

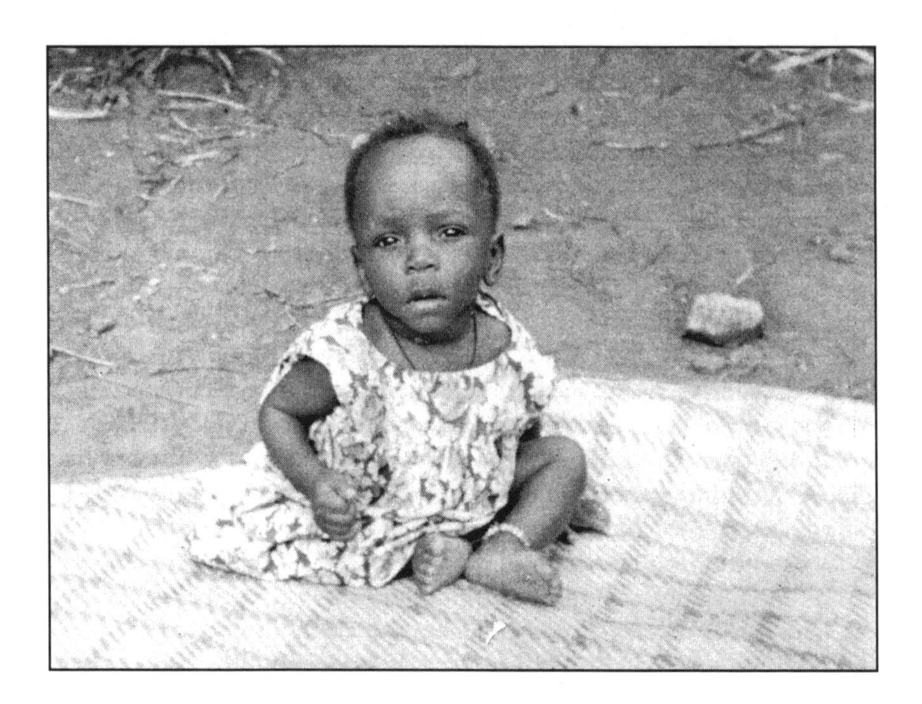

FIGURE 2.5. Petero at thirty-four weeks. *Source:* Ainsworth, *Infancy in Uganda*, 301, image 31.

growing popularity of bottle feeding in .. The first to raise concern over generalized bottle feeding in the region was Dr. Welbourn. Through her mobile child welfare clinics she was well placed to observe changing patterns in infant feeding and the record cards kept for each child provided data for later analysis.[87] According to Welbourn, most mothers in Uganda exclusively breast-fed until the mid-1950s, when she noted that "it has now become a common sight in Uganda to see an African baby being fed . . . with various mixtures of milk, cereal or tea from a bottle."[88] In comparing the records from 1955 with those taken in 1950, Welbourn found evidence of a distinct shift toward supplementary bottle feeding. Although most of the children were still breast-fed, the percentage of infants under one and six months of age who were also bottle fed more than doubled in just six years. Ainsworth also reported that "except for one baby, supplementary feeding before three months of age consisted entirely of milk given by means of a nursing bottle and excluded solid foods. Bottle feeding of very young babies

FIGURE 2.6. Waswa and Nakato at sixteen and a half months and their mother. *Source:* Ainsworth, *Infancy in Uganda,* 306, image 33. © 1967 The Johns Hopkins Press. Reprinted with permission of Johns Hopkins University Press.

was observed in even the least acculturated families. . . . The supplementary bottle feeding often consisted of weak tea mixed with powdered milk and sugar."[89] Furthermore, Welbourn's figures indicated that "supplementary feeding was nearly always followed by early weaning."[90] Infants given supplementary bottles were weaned, on average, six months earlier than those who were breast-fed alone.

This trend to supplement breast feeding and wean early was also emphasized by Dr. Latimer Musoke in his analysis of the pediatric admissions at Mulago in 1959. Like Welbourn, Musoke observed that among "the Baganda, although many do still breast-feed their babies, early supplementary feeding is increasing."[91] In a study of weaning practices in Buganda in the mid-1960s, Namboze also found that among mothers residing in Kampala or the surrounding area, nearly half weaned their infants before their first birthday and 34 percent gave their young children formula, milk, or cereal "instead of breast milk."[92] Moreover, Welbourn's clinic record cards, five years following her first assessment, revealed that bottle feeding had risen a further 20 percent to the point that "in 1960, 42 percent of babies attending child welfare clinics were being given supplementary bottle feeds before the age of 6 months."[93]

Based on a widely held belief that "highly educated" women were capable of bottle feeding successfully, Welbourn even explicitly encouraged bottle feeding among some parents. In Ainsworth's study, many of the mothers who used bottles to supplement their child's diet did so on the advice of the health clinic, and, in one instance, it was observed that Welbourn "recommended" the wife of a hospital clerk give her son "some of his milk from a nursing bottle."[94] Many of my informants confirmed that this was common. Catherine Nansamba, who was a young mother in this period, explained that the medical personnel at the rural health center near her home taught that bottle feeding was advantageous if used correctly and only a problem when mothers lacked knowledge of proper use.[95]

In light of the hazards of bottle feeding in regions with insufficient supplies of safe drinking water, biomedical personnel characterized the growing popularity of bottle feeding in alarmist terms.[96] Among the many dangers for infant health was the fact that even supplementary bottle feeding "tended to replace breast feeds," and as a mother's supply of breast milk is directly determined by the demand of the suckling infant, less frequent nursing quickly translated into an often irreversible reduction of breast milk.[97] In addition, the expense of bottle feeding with store-bought milk or imported infant formulas was far beyond the means of the vast majority of Ugandans. According to Jelliffe, "to bottle-feed a four-month old baby on a reputable full cream dried milk costs the equivalent of between one-third and one-quarter of the total salary of an African worker."[98] Michael Latham put the figures for Tanzania in the mid-1960s at 51 percent and found in Kenya a decade later that a full 58 percent of the minimum wage was required to adequately feed a four-month-old child with milk or infant formula.[99] Despite the very real issue of affordability, many observers at the time focused instead on the supposition that women diluted the milk or formula because they believed that expense indicated power and potency, shifting the emphasis from questions of economic considerations to perceptions of individual and cultural ignorance.

The result was that bottle-fed babies faced a significantly higher risk of consuming an insufficient diet. Not long after mothers began supplementary bottle feeding, Welbourn found that "most of the children showed signs of undernutrition."[100] This was certainly the case for many of the children documented by Ainsworth. Waswa was so severely underweight that by current standards he would be classified as "severely malnourished"; his sister Nakato, Magalita, and Petero were likewise "moderately malnourished" and underweight by contemporary standards; and Suliamani, who had been thriving, was at 8 months also approaching a moderately malnourished state. In fact, the bottle-fed babies in Welbourn's sample were an average of one to two pounds lighter than those who were exclusively breast-fed and two of the three children who became severely malnourished while attending Welbourn's clinics in 1955 were reportedly being fed with a bottle.[101] Moreover, in very young children struggling with undernutrition, an average weight differential of between one and two pounds was significant.

Problems of undernutrition were exacerbated by the far greater frequency of infectious disease in bottle-fed infants. The prevalence of waterborne pathogens exposed them to a host of infections that they might otherwise avoid and the consequent reduction or absence of breast milk in the diet meant that these very same children were not acquiring the antibodies that passed from mother to child, providing breast-fed babies with some protection from infection. Welbourn's evidence indicated that the children who were fed supplementary bottles were twice as likely to develop diarrhea, vomiting, and acute respiratory infections.[102] The illnesses typically associated with bottle feeding were also clearly evident in the pediatric wards of Uganda's central hospital by the close of the 1950s, and physicians were well aware that the growing prevalence of gastroenteritis among Mulago's pediatric patients represented a shift from earlier disease patterns.[103] Musoke wrote for instance that "it is significant that a disease which seven years ago was considered to be quite uncommon, now tops the list of admissions. Davies . . . stated that [in 1950] diarrhoea and vomiting were relatively unimportant causes for admission of children in Mulago Hospital. . . . The present analysis shows that the diarrhoea and vomiting syndrome has become one of the major problems of the ward." Moreover, biomedical practitioners rightly attributed this epidemiological shift to the rise of supplementary bottle feeding; as Musoke pointed out, exclusively breast-fed children made up less than 14 percent of the cases admitted for diarrhea and vomiting.[104]

The connection between the distribution of dried skim milk, skim milk–based kwashiorkor therapy, and the growing use of nursing bottles in Uganda was not at first appreciated. Despite the widespread recognition that bottle feeding was on the rise, biomedical personnel uncritically accepted the view that bottle feeding was an inevitable result of urbanization, westernization, and civilization. Trowell, in the first edition of the classic text *Diseases of Children in the Subtropics and Tropics*, opened the infant feeding chapter with a treatise on the "natural" progression from breast to bottle feeding: "The incidence and duration of breast feeding reflect the customs, social patterns, economic development and sophistication current in any community. Perhaps those who are nearest to nature are themselves most natural. . . . Under the impact of civilization and sophistication the movement at first is always away from breast feeding to bottle feeding."[105] Trapped within this teleological framework, biomedical personnel failed to register the role of their own therapeutic and preventive practices. In fact, it was not until the late 1970s that Jelliffe and his wife regretfully acknowledged that "it was a nutritional tragedy that the well-intentioned, widespread feeding programmes in developing countries in the 1940s and 1950s should have been primarily concerned with the distribution of dry skimmed-milk powder. . . . As far as mothers of young children were concerned, this can only have appeared as endorsement of bottle feeding, with a resulting displacement effect on breast-feeding."[106]

Seeking a single-origin explanation of bottle feeding trends in Uganda during the mid-twentieth century risks oversimplifying the complex myriad of factors that undoubtedly influenced shifting infant feeding patterns. The emphasis placed on urbanization and modernity in Uganda not only obscured the role of kwashiorkor treatment and prevention, it also inaccurately reflected the realities of social and economic change during the period of skim milk distribution and bottle feeding. The expansion of urban centers in Uganda at this time was minimal and did little to alter the fact that Uganda remained predominantly rural. Thus, while some observers spoke of "a rapid growth of towns" and took note of the "large influx of population into

Kampala," estimates suggest that as little as 3 percent of the total population of 6.5 million lived in Kampala or one of the other four urban centers that contained more than 10,000 inhabitants.[107] In 1961, it was still said that "relatively few Africans have come to live in urban areas. In marked contrast to the West African pattern, the people of Uganda live neither in villages nor towns."[108]

Uganda remained decidedly rural due in large part to the fact that cash crop production continued to be the engine of economic growth and commercial expansion. With few alternative avenues of upward mobility, a "middle class" of entrepreneurial farmers emerged to take advantage of the postwar boom in world prices for agricultural products like cotton and coffee.[109] A steady flow of immigrant laborers from Rwanda and Burundi as well as the fragmentation and sale of land allowed small-scale farmers to expand their holdings and begin farming on a more commercial basis.[110] In the absence of a mining or plantation sector, Ugandan farmers had little incentive to put down their hoes and flock to town to seek a better life. Instead capital accumulation involved an expansion of rural holdings, not migration away from rural areas. Moreover, a long history of rural-urban integration and mobility around the royal and administrative capitals meant the urban populations that did exist were far from permanent. More often than not, as one observer noted, "the capital-dweller maintained a foothold in the countryside—usually a holding of land from which he might draw supplies of food."[111] With women and children especially rooted in rural homesteads, physicians and scientists who worked closely with severely malnourished children at Mulago repeatedly acknowledged that "most of [their] patients . . . were absolutely rural people, living out in the shamba, growing all their food, living the traditional Baganda life."[112] Sustaining the fiction that bottle feeding could be attributed solely to an inevitable process of urbanization and the march of modernity entailed arguing, as Jelliffe and his wife later did, that "pressures towards urban life-styles are no longer confined to the cities, but are tending to spread out to rural areas in some countries by word of mouth, and particularly through the channel of the transistorized radio. This process of 'mental urbanization' has already tended to affect ways of life, including the ability to breast-feed in some rural parts of the world."[113]

Two developments in the early 1970s further impeded awareness of how the medicalization of malnutrition and the emphasis on skim milk in both therapy and cure contributed to the rise of bottle feeding in Uganda. Just as the notorious Ugandan dictator Idi Amin began his reign of terror, the global focus on severe protein deficiency became the subject of significant controversy and critique. In 1974, Dr. Donald S. McLaren published an article in the preeminent medical journal the *Lancet* entitled "The Great Protein Fiasco."[114] McLaren criticized the emphasis placed on severe protein malnutrition by international agencies linked to the United Nations and effectively undermined the widespread view that there was an "impending protein crisis" of global proportions.[115] His article triggered a decisive shift away from protein malnutrition as the main focus of efforts to improve nutritional health in young children. The pendulum swung so swiftly in the opposite direction that kwashiorkor research and programming ground to a virtual halt. Those who championed McLaren's critique represented a growing group of biomedical practitioners who, like McLaren, had been working in regions of the world where protein deficiency did not appear to be the most prevalent form of childhood malnutrition. For McLaren and many others, it was marasmus or undernutrition that warranted greater attention but had been overshadowed by the international preoccupation with protein. McLaren's intention had not been to question the science implicating protein deficiency as the cause of kwashiorkor, merely the general application of that evidence in the international arena. The interpretive error, as McLaren repeatedly argued, was "the extrapolation by others of the atypical experience in Africa to the rest of the developing world."[116] McLaren even explained in a letter to Cicely Williams that it was not the evidence of protein malnutrition in Africa that was problematic, merely the extrapolation of that evidence onto a global scale.[117]

Regardless of his intentions, McLaren's critique dealt a final blow to the protein paradigm. Kwashiorkor and marasmus were increasingly understood as merely opposing ends along a spectrum of childhood malnutrition and instead of protein deficiency, physicians and scientists spoke of protein-calorie (PCM) and protein-energy malnutrition (PEM). Yet in defining the problem as a misplaced emphasis upon the wrong nutrient, McLaren's "Great Protein Fiasco" represented a missed opportunity to critically examine the unintended consequences of narrow, single-causative conceptions of poor nutritional health and disease. Rather than subject the highly targeted magic-bullet approach of the high-protein food program and its failure to the kind of scrutiny that could provide insights for future efforts, the article made it an embarrassment to be swept under the rug and forgotten. Additionally, assuming that the greater problem of marasmus had simply been overlooked preserved an image of global malnutrition as a static and unchanging problem amenable to straightforward scientific solutions. This view missed the shifting epidemiology of childhood malnutrition that was clearly evident in Uganda. McLaren concluded his indictment of international efforts to combat the so-called worldwide protein crisis by insisting that the true costs be measured in terms of the children "lost in the unchecked scourge of malnutrition."[118] To this we must add many of the children who died as a result of the bottle feeding that was unintentionally encouraged by the failure of the high-protein food program and the distribution of dried skim milk at hospitals and health centers.

It was also in the early 1970s that scandal broke over the global marketing of infant formula. The controversy began at a UN Protein Advisory Group conference in 1970 when Jelliffe launched a vociferous critique of the commercial marketing of infant formulas and the disastrous consequences for infant health and welfare.[119] While the PAG worked with industrial partners to develop marketing guidelines that could preserve profits without directly promoting bottle feeding, Jelliffe argued for a more aggressive approach.[120] In using the phrase "commerciogenic malnutrition," Jelliffe suggested that the starvation, diarrhea, and death associated with marasmus was often "caused by the thoughtless promotion of these milks and infant foods."[121] Then in 1973 the *New Internationalist* published an article entitled "The Baby Food Tragedy" and the following year a charitable organization published a report with the inflammatory title "The Baby Killer."[122] Mike Muller, author of "The Baby Killer," described the organization War on Want as "one of many Third World charities indebted to certain of the milk companies for donations of infant foods and other products of relief programmes. In disasters and other *abnormal* situations, these products can be extremely useful," but his report went on to accuse the baby food industry of unscrupulously promoting and advertising them in order to convince mothers to choose bottle over breastfeeding.[123] In addition to providing free samples and gifts, and promoting breast milk alternatives in hospitals and health centers, infant formula manufacturers reportedly peddled their products through "sales girls dressed in nurses' uniforms."[124]

A highly contentious international public campaign ensued. Citing remarkable declines in breastfeeding and the alarming malnutrition and infant mortality resulting from diluted and contaminated bottle feeding in impoverished world regions, activists launched boycotts and demonstrations, notably targeting the Swiss multinational company Nestlé. Following a congressional hearing in the United States and numerous international meetings, the World Health Assembly adopted the WHO International Code of Marketing Breast Milk Substitutes in May 1981.[125] The infant formula controversy succeeded in directing international attention to the connection between bottle feeding, marasmus, and infant mortality. Activism around the issue highlighted the nefarious activities of companies who put profits before public health and established an unprecedented set of regulations that aimed to prohibit such practices. But it also inadvertently further diverted attention from the role that medicalizing malnutrition had in the rise of bottle feeding and growing problem of undernutrition and marasmus. This diversion also inadvertently served to maintain the fiction that a narrow biomedical panacea could solve the problem of severe malnutrition—a fiction that unfortunately persists to this day.

The rise of the protein paradigm grew out of evidence implicating protein as the cause of severe acute malnutrition. The therapeutic efficacy of Dean's high-protein therapy together with pathological indications of protein deficiency in Ugandan children dying of the condition propelled this single nutrient onto the international stage at the very moment that the newly formed WHO and FAO created an Expert Committee on Nutrition. With protein malnutrition at the forefront of international health, physicians and scientists working in Uganda were influential in the evolution of global endeavors to treat and prevent the condition. The emerging biomedical discourse constructed the problem of protein deficiency as an endemic problem in resource-poor settings. The assumption was that kwashiorkor had long, and perhaps always, been an ongoing and recurrent condition among people living in particular global environments.

Experts then set to work developing high-protein weaning foods that could prevent severe acute malnutrition and close the global "protein gap." While awaiting the rollout of the high-protein food program, emergency measures focused on the distribution of skim-milk powder. The ready availability of powdered milk in the postwar period and the success in treating severe malnutrition with skim milk–based formulas made it appear to be an ideal source of dietary protein for the immediate prevention of severe malnutrition in young children. The eventual failure of the high-protein food program meant skim-milk powder distribution continued for at least two decades—moving from an emergency measure to the essence of prevention in Uganda and other parts of the world. With skim milk as the basis of both prevention and cure, and in light of the remarkable therapeutic efficacy of skim milk, this medicalization of malnutrition encouraged bottle feeding with deleterious consequences for child health and nutrition.

In contrast with the more well-known controversies over the global marketing of infant formulas by multinational corporations, biomedical personnel unintentionally endorsed bottle feeding as a result of well-intentioned but ill-conceived efforts to save young lives. Unlike the pronatalist policies of colonial administrators in the Belgian Congo who promoted skim-milk powder in order to explicitly replace breastfeeding, reduce birth-spacing, and thereby increase population and labor supplies,[126] physicians and scientists advancing dried skimmed milk as therapy and prevention sought to encourage rather than discourage breastfeeding. Believing that bottle feeding was an inevitable result of urbanization and the civilizing, Western influences of modernity, concern over the growing popularity of nursing bottles did not immediately prompt scrutiny of skim milk distribution. This teleological view missed how Ugandans appropriated the efficacy of kwashiorkor therapy, incorporating a new category of illness—olbuwadde bw'eccupa—into the local lexicon of misfortune and disease, and so it also missed how their engagement with this initiative shaped the outcome. The power of skim milk in the treatment of eccupa converged with distribution at hospitals and health centers to undermine confidence in breastfeeding and, as a result, further compromise child health and nutrition in Uganda.

3 THE MIRACLE OF *KITOBERO*

Throughout the 1950s, experts working to treat and prevent severe acute malnutrition remained convinced that it was a straightforward matter of devising high-protein weaning foods that could fill a worldwide gap in available protein. By the early 1960s, confidence in this targeted, biomedical approach began to wane. The winds of change that saw many African colonial territories achieve independence (including Uganda in 1962) also bore witness to substantial shifts in the science and provision of medicine. In addition to contributing to the rise of bottle feeding, the medicalization of malnutrition impeded longstanding efforts to prevent severe acute malnutrition in Uganda. Local demand for a treatment that worked meant that malnutrition became a disease requiring hospital treatment. Faced with the failure of prevention, biomedical personnel evaluated their approach and shifted tack. What emerged was a hybrid, a synthesis that built on the biomedical infrastructure and expertise in Uganda *and* on Ugandan engagements with biomedical treatment and care. The result was also a hybrid in that it brought together pioneering approaches to health promotion from disparate parts of the world, but did so *in Uganda*, as a local response to local obstacles to health promotion. This chapter examines the establishment of a comprehensive public health approach to the prevention of severe acute malnutrition and argues that the lasting legacy of the program in one region of Uganda points to the advantages of hybrid public health endeavors that build local capacity, sovereignty, and health services sustainability—in short, local public health programming masquerading as global health.

From Relapse to Rehabilitation

Just prior to his death in 1964, Dean reportedly warned the physicians and scientists working to combat severe acute malnutrition in Uganda that they were "in danger of knowing more about malnutrition and less about its elimination than anywhere else in the world."[1] After over a decade of trying to devise a protein-rich weaning food that could prevent the condition,

Dean and his colleagues had to face the fact that their efforts had been an abject failure. What brought the realization of failure home was the problem of relapse.[2] Children previously treated for severe acute malnutrition and then brought back to the hospital for the same life-saving treatment became a particularly poignant indication of the failure of prevention. Saving the lives of severely malnourished children a second or third time signaled the inadequacy of their initiatives. Relapse brought the medicalization of malnutrition into sharp relief and prompted physicians to critically reflect on their approach. As one physician later explained, "The free distribution of DSM [dried skimmed milk] through medical centres would, it was hoped, help to remove malnutrition. Ten years of distribution finds the incidence almost unchanged. Paradoxically those who have made the best use of it are often sadly reliant on it and have made no other change in their children's feeding. Consider the comment of one such mother, with her child developing kwashiorkor for the second time, 'Doctor, what could I do; the milk ran out?'"[3]

What medical personnel began to question was the value of the curative framework of hospital medicine. Initially, biomedical practitioners viewed hospital treatment as an ideal opportunity to impart essential health education. A Sudanese physician, for example, observed at a 1961 health education seminar that

> Children admitted to hospitals are accompanied by their mothers who stay beside their beds. . . . This system offers a vital chance for health education, especially in nutritional disorders. The cause of the *disease* is explained to the mother who is advised to watch how the child is fed and cleaned in hospital so as to follow the same method at home. This proved to be fruitful in kwashiorkor. . . . The mother in hospital is shown how milk is prepared and is encouraged to help the nurses in feeding her child.[4]

But in the context of relapse, the presumed benefits of the hospital as a fruitful site for prevention were becoming increasingly unclear. Rather than demonstrate that severe acute malnutrition was a preventable condition, the extremely disciplined system of prescribing and delivering the therapeutic formula combined with routine injections of antibiotics, chloroquine, and other medications suggested that severely malnourished children suffered from a *disease* requiring elaborate curative interventions.[5] Physicians at Mulago therefore began to realize the futility of teaching parents and guardians that the condition could easily be prevented as their children underwent hospital treatment. As one pediatrician lamented, "We were teaching in the wards with everything curative surrounding the mothers."[6]

Biomedical personnel also reasoned that in the highly curative framework of the hospital ward, mothers became passive observers of biomedical interventions. Relapse was read as an indication that curative therapies fostered a growing reliance on clinical medicine.

> With the pressure on hospital beds, such as obtains at Mulago Hospital, only those children with severe or complicated kwashiorkor are likely to be admitted. For them the dramatic medical procedures such as feeding by a tube into the stomach or into a vein and courses of injections are often needed. These procedures in the bewildering surroundings of a modern hospital are accepted passively by the mother who, delighted at the cure of her child, goes home with her misconception about kwashiorkor subtly confirmed by the medical treatment. With child feeding methods at home unchanged, the recurrence rate is naturally high.[7]

However, positing the problem of relapse as a passive reliance misses or misconstrues an essential part of this history of nutrition in Uganda. The fact that people interpreted and incorporated the development of a novel, effective, and highly curative therapy as evidence of a new disease suggests a particular, and a particularly active, engagement with biomedicine. In the mid-twentieth century, therapeutic networks of parents and relatives brought severely malnourished children from ever greater distances to Mulago Hospital for treatment because their children suffered from what appeared to be a disease that could only be treated in the hospital. These parents and guardians sought and demanded hospital therapy for their children and this demand, at times a repeated demand, was among the consequences of the medicalization of malnutrition in Uganda. On the ground, this medicalization transformed severe acute malnutrition into a *disease*, a disease that required extensive, even emergency, clinical treatment. This local engagement with and response to the development of effective therapeutic measures was pivotal to the failure of prevention and to the subsequent decision to try an entirely new approach. Africa's first nutrition rehabilitation program drew its inspiration from a number of distinct places, but first among them were the circumstances encountered in Uganda—circumstances that were shaped largely by the changing local engagement with the nutrition research and programming. As diagnostic uncertainty morphed into the medicalization of malnutrition, local hesitation became local demand and prevention became elusive.

But the times were also changing. The mid- to late 1950s saw the rise of preventive social medicine and community-oriented health models in Uganda. The long history of advanced medical training and research made the newly independent nation an important node in networks of international health experts engaged in these novel approaches to problems like malnutrition. In particular, Uganda became a destination for those fleeing the worsening conditions of apartheid South Africa, and the cross-fertilization between those who had pioneered community-oriented health promotion models at rural health centers in South Africa and the growing number of pediatric experts and those working specifically on malnutrition paved the way for the advent of a novel public health approach. This cross-fertilization first bore fruit at a rural health center in the village of Kasangati, a short distance from the capital city and medical school. It was established in 1959 as a rural training facility under the auspices of Makerere's Department of Preventive Medicine. The central philosophy behind the work at Kasangati was to provide future doctors with practical training in a rural setting that integrated prevention with curative medicine. In addition to emphasizing social facets of health and the importance of circumstances within the home, the approach at Kasangati stressed the need to "consider the whole patient in relation to his environment and not merely . . . the illness or diseased organ."[8]

Moreover, the health center provided cutting-edge services, and the benefits of the services provided at Kasangati drew people to the unit. Kasangati was the first to offer family planning and immunization. According to Josephine Namboze, who was both the first medical officer in charge of Kasangati and the first female doctor in East Africa, in December 1962 when they began giving children vaccines to prevent a number of diseases, including measles, polio, diphtheria, and pertussis, such childhood vaccines were only available to expatriate children at the mission-run Mengo Hospital in Kampala.[9] In fact, the nurses initially refused to administer the vaccines for fear of retribution in the case of an adverse reaction, and until the advantages of immunization became clear to the surrounding community, Namboze had to give the vaccines herself.[10]

Efforts to ensure accessibility in Kasangati included mobile children's clinics designed (and in some cases requested by the residents) to minimize how far parents of young children had to travel, ideally to no more than a mile, and these clinics boasted attendance rates of 75 percent.[11] According to one observer, the number of patient visits increased threefold over a four-month period, and, in his view, "the obvious satisfaction on the part of the villagers with the unit is everywhere apparent."[12] This evidence of local approval was also due to the high value placed on community input and participation, which, according to Namboze, meant that the health center explicitly avoided a top-down approach. In an interview, Namboze used the example of nutrition education to illustrate how this worked. According to Namboze, "mothers got so knowledgeable that we asked them to actually give the health education themselves. . . . It was not one way, people in the community were also participating and they would see the advantages of these services, and the knowledge, and they would start teaching one another, and health in the community started improving."[13]

In addition to those who came from South Africa, Uganda also drew experts from further afield, including Derrick B. Jelliffe, who was appointed as the UNICEF (and first full) Professor of Pediatrics and Child Health at Makerere. Jelliffe arrived in East Africa not as a young colonial medical officer, but as an established international expert in child health and nutrition. Once in Uganda, he joined forces with a group of like-minded experts emphasizing the social and cultural facets of nutritional health and its promotion.[14] In 1961, for example, Jelliffe and his colleagues organized a conference on custom and child health in Buganda with the explicit aim of identifying customary practices that could be the basis of preventive measures—an approach representing a distinct shift from the prevailing colonial framework in which cultural practices were evaluated solely in an effort to curb those held responsible for poor nutrition.[15] Over the course of several years, Jelliffe and physicians working in Makerere University's newly established Department of Preventive Medicine formed a team of African and expatriate personnel to conduct a series of community child health studies assessing the prevalence and underlying causes of malnutrition in several disparate districts of Uganda.[16] In attending to what they called the "ecology" of malnutrition and providing crucial epidemiological evidence, particularly in areas outside Buganda, this work also signaled a shift in medical thinking and significantly expanded the already formidable body of knowledge concerning childhood malnutrition in the region.

In the early 1960s Jelliffe began assembling a group of pediatricians at Makerere, including an Australian physician, Ian Schneideman, and Paget Stanfield, who began working in Uganda in 1962 through a link between

Makerere and the Great Ormond Street Sick Children's Hospital.[17] It was also Jelliffe who first enlisted Josephine Namboze. After graduating from Makerere, Namboze briefly worked with Jelliffe in pediatrics and it was there that, as she put it, she "developed an interest in community work or preventive and promotive" medicine. While treating children in the pediatric wards, Namboze began to appreciate that their repeated illnesses were preventable: "The cases I was dealing with were repetitive . . . preventable at a community level. I realized that my staying there . . . wouldn't solve a problem. I felt that I had to work with the community and with the families to be able to prevent these problems."[18] Jelliffe then arranged for Namboze to pursue advanced training abroad, and when she returned, she applied her passion and expertise to the health problems encountered in and around Kasangati. Namboze's conclusion that a different approach was necessary to improve the health and well-being of the Ugandan people parallels the decision that pediatricians encountering relapse among severely malnourished patients made to pioneer a new public health model that brought the insights behind the work at Kasangati (which originated in South Africa) together with the concept of nutrition rehabilitation (which stemmed from similar endeavors in South and Central America). And yet core aspects of the resulting program were drawn from the local culture and heritage of the Ugandan people.

In February 1965, several wards of the old Mulago Hospital reopened as the "Save the Children Fund Nutrition Rehabilitation Unit."[19] The concept behind nutrition rehabilitation was "to bridge the ill-defined borderland between classical 'treatment' and 'prevention'" by integrating nutrition education with the rehabilitation of malnourished children.[20] From 1965 onward, children brought to Mulago's Government Dispensary (or one of the hospitals and dispensaries within a sixty-five mile radius) suffering from malnutrition but not requiring immediate hospitalization and emergency services were referred to the Nutrition Rehabilitation Unit. These children were either admitted as outpatients or, together with their mothers or female guardians, as inpatients in the unit's residential ward. Children whose mothers or guardians did not seem interested or receptive to the nutrition rehabilitation program and those requiring extensive and immediate therapeutic measures continued to receive treatment in the main hospital.[21] This provision guaranteed that the children sent to the Nutrition Rehabilitation Unit could fully recover from food and milk prepared and administered exclusively by their mothers and attending relatives (see figs. 3.1 and 3.2).

Moreover, the costs were a fraction of those for hospital treatment. The average cost for those who attended the unit on an outpatient basis was found to be roughly 34 shillings (Shs.), and the cost for those admitted as inpatients came to approximately Shs. 550 for six weeks or Shs. 13 per day. By comparison, hospital treatment for two weeks came to around Shs. 840.[22] In keeping curative treatment at the unit to an absolute minimum, the program expanded the role of mothers while visibly diminishing the role of biomedical personnel. Their seemingly exclusive role empowered mothers in the prevention of malnutrition in the future.[23] Although the program appeared to focus on the rehabilitation of severely malnourished children, the central aim was to educate and train mothers. Mothers at the unit were therefore "learning by doing," which became one of the core pedagogical principles of the program.[24]

The active role that mothers had in all aspects of the rehabilitation process was also crucial to what emerged as the second key philosophy of nutrition rehabilitation in Uganda: "seeing is believing."[25] In the first decade of the program, at any given time, approximately twenty mothers and their children were staying at the unit for a period ranging from three to eight weeks.[26] It was soon realized that admitting and discharging women and their children according to their individual needs and circumstances, and on a staggered basis, ensured that the children at the unit were at different stages in their rehabilitation. Those who were well on their way to recovery could, it was discovered, illustrate for others how a child's nutritional health was restored by their mother's hand and through the provision of food alone. Women and attending relatives not only watched their own children but learned by observing their peers and their children as they also visibly recovered from what was once a fatal condition. Mike Church, the medical officer in charge of the unit, noted that "the radical change in three to four weeks of a child with kwashiorkor on treatment gives invaluable credibility, so needed in nutrition education."[27] What these women observed at the unit differed substantially from the treatment previously witnessed in Mulago's pediatric wards.

In a move to demedicalize malnutrition, the program sought to fundamentally divorce or distance nutrition education from the emergency measures of the hospital. The primary instructors at the unit were, as a result, not biomedical personnel but mothers of recovering children. "Mothers whose children are cured," wrote Stanfield, "act as teachers, and their healthy children are living proofs of their teaching."[28] This was particularly

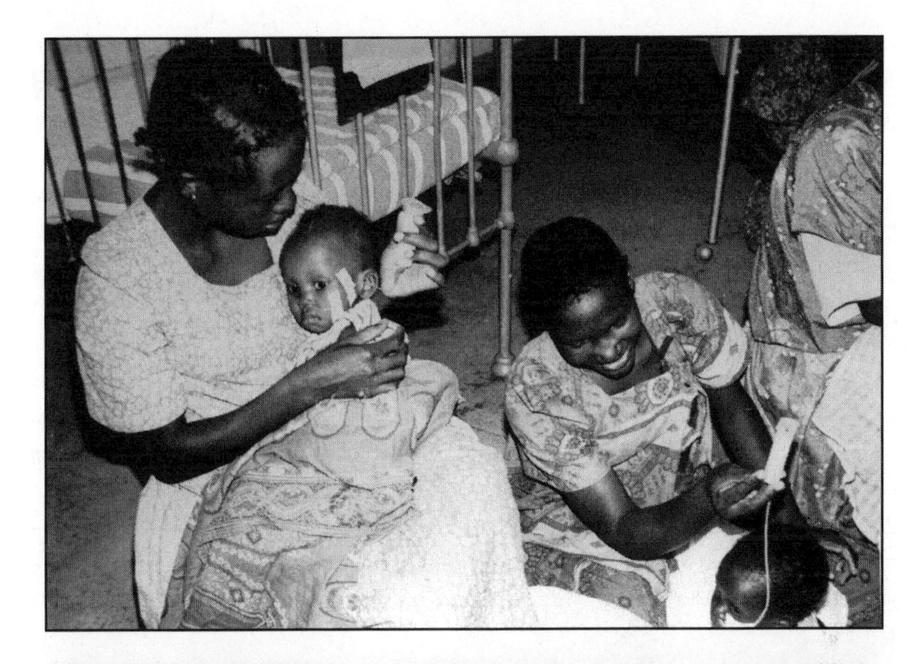

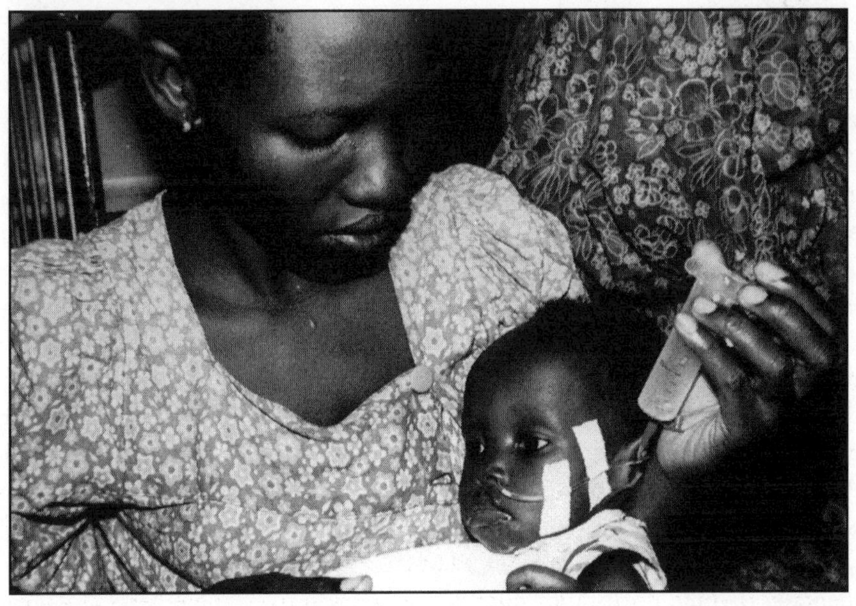

FIGURES 3.1 AND 3.2. Women delivering therapeutic formula to malnourished children at Mwanamugimu, c. 1965. Courtesy of Paget Stanfield.

true in the mid- to late 1960s, as the vast majority of the parents and children served by the unit (as many as 90 percent) participated in the program on an outpatient basis at daily clinics. Whereas the residential unit annually admitted approximately 150 mothers and their children, 1,500–2,000 malnourished children were routinely seen at that outpatient clinic for the first time each year, and because parents brought children to the unit's outpatient department an average of four times, the total number of outpatient visits ranged, in this period, from over 8,000 to nearly 9,000 annually. Efforts were also increasingly made to maximize the impact of mothers who were veterans of the program by scheduling follow-up visits so that the returning parents far outnumbered (and thereby increased their influence on) parents who were bringing their children to the unit for the first time.[29] The mothers who resided at the unit as inpatients were also integral to the outpatient teaching program. They became experts not only in the preparation of nutritious foods but in the instruction of their peers, and these skills were on display as they prepared demonstration meals for those attending the outpatient clinic (see fig. 3.3 and 3.4).[30]

The facilities used to house the unit were also designed to create a less curative environment, one as distinct from the hospital as possible. The far more rudimentary, one-story buildings of the old Mulago Hospital stood in sharp contrast to the newly constructed general teaching hospital built at the base of Mulago Hill. In fact, the construction of the new Mulago Hospital was viewed by many of the resident physicians as such a retrograde step that they referred to the thousand-bed facility as the "Triassic Palace." According to Stanfield and Church, this final "gift" of the British colonial government, which opened in the same year that Uganda achieved independence, was designed by British architects according to the prevalence and history of disease in England. As a result, the pediatric wards were practically forgotten and so inadequate that they quickly became overcrowded. Less than a decade after opening, as Church recalled, "children were on the floor, they were under the beds, they were in the corridors." Moreover, the wards of the new hospital were part of an extensive multistory building, which opened onto the hallways and corridors of the main hospital and did not, therefore, permit parents to move freely from the ward out onto the grass, where they had formerly cooked and cared for their sick children.[31] Using the wards of the old Mulago Hospital and building on the greater accessibility of the single-story buildings, the Nutrition Rehabilitation Unit was, as intended, a far less biomedically driven environment that was centered

FIGURES 3.3 AND 3.4. Cooking demonstrations at Mwanamugimu. Courtesy of Paget Stanfield.

around mothers and their children. Efforts were increasingly made to incorporate aspects of life from the average Ganda home. Replicating home life was understood as crucial to the efficacy of the program, which was thought to hinge on the widespread adoption of only two changes in infant feeding: to measure and mix a young child's food. The implementation of these changes in feeding practices was seen as feasible only if they were kept to an absolute minimum. Thus all elements of the program explicitly sought to encourage only slight adjustments in the way that Ganda fed their children. As Ian Schneideman explained, mothers at the unit prepared the therapeutic diet using "cheap locally available traditional food and with cooking facilities to which she is accustomed," which, as he emphasized, meant "in the customary way with only slight modification" and with "materials and utensils that are available in every home."[32] Creating an environment that was reminiscent of the average Ganda home was seen as so important that an attempt was even made to build demonstration kitchens with mud floors.[33] In the two demonstration kitchens, mothers steamed food over an open fire after it had been wrapped in banana leaves and placed in large pots. In fact, the main distinction between the design of the unit kitchens and the average Ganda hearth, aside from the fact that the kitchens were open facilities suitable for large-scale cooking demonstrations, was that the fires were raised off the ground in order to prevent young children from inadvertently getting burned. The presence of a demonstration vegetable garden, compost, rabbit hutch, poultry coop, and pit latrine contributed to the practical public health education offered at the unit and further distinguished the program from the far more curative framework of the neighboring hospital (see fig. 3.5).

Great importance was also placed on the composition of the staff. The importance of demedicalizing malnutrition made it imperative that the majority of the personnel involved in the day-to-day operations and the ongoing evolution of the program were not, first and foremost, biomedical personnel. The woman chosen as the first director of the Nutrition Rehabilitation Unit was Gladys Stokes (neé Nansubuga), a local schoolteacher specializing in nutrition at Gayaza High School (see fig. 3.6). Stokes directed a staff that included a medical assistant, a newly trained nurse, and a pediatrician, but the majority of the personnel were explicitly "untrained."[34] In fact, de-emphasizing biomedicine at the unit was so essential that even though the unit was important to medical training, doctors and medical students were not permitted to enter the unit grounds wearing their white coats or stethoscopes. As Stanfield and Church recalled, the overarching policy was

FIGURE 3.5. Mwanamugimu Nutrition Rehabilitation Unit as seen from original entrance. Courtesy of Paget Stanfield.

FIGURE 3.6. Staff of Mwanamugimu Nutrition Rehabilitation Unit, c. 1965. Courtesy of Paget Stanfield.

unequivocally designed "to keep doctors out."[35] In fact, Stanfield later wrote that the "methodology of prevention, has led those involved in this problem to break out of the confines of the hospital situation. . . . The doctor is realising that he holds only a very small part of the total answer to malnutrition."[36]

It was also considered crucial that the staff be "locally appointed." As the unit was opened only a few years after political independence, the importance placed on enlisting an entirely African staff was both influenced by and went far beyond the political imperatives of Africanization. Rather, the concept of Africanization spoke to the very core of the program itself, as the expatriate physicians believed that its success hinged on whether or not it became a Ugandan program. They viewed themselves as temporary facilitators in the establishment of a permanent infrastructure. By providing a foundation for the content, philosophy, and future direction of the program to emerge and evolve, they hoped to create a program that would best meet the changing needs of the Ugandan people. In a description of the program more than a decade after its inauguration, Stanfield observed that "the solution to a problem very often lies within the rich tradition of the culture if it can only be tapped by local people who can live in two cultural frameworks."[37] The Nutrition Rehabilitation Program aimed to succeed where previous programs of prevention had failed by employing a staff of African women and men who were not thoroughly ensconced in the epistemology of biomedicine. Africanization was, therefore, a fundamental step in the establishment of a novel, hybrid public health program that sought to prevent the incidence of severe acute malnutrition by demedicalizing malnutrition and empowering mothers and Ugandan communities in the management and promotion of nutritional health.

Kitobero—*A Synthesis*

It did not take long for Stokes and her team to begin molding the Nutrition Rehabilitation Program into a lasting public health initiative. Among the more immediate changes was the name of the unit. At the suggestion of Stokes and her fellow staff, the unit was renamed the Mwanamugimu Nutrition Rehabilitation Unit after the well-known Luganda proverb, *Mwanamugimu ava ku ngozi.* Often translated as "A healthy child comes from a healthy mother," the proverb appeared to capture the essence of a program aimed at improving child health by empowering mothers.[38] Her team also immediately set to work transforming nutrition education at the unit through a unique synthesis of local and biomedical ideas about food, health, and nutrition. Although physicians in Uganda were already becoming aware of the gap between their conceptions of the basic food groups and local ideas surrounding food and hunger, according to Stanfield and Church, "it took someone like Gladys Stokes to say, 'Look, you haven't got the faintest idea how to communicate with these mothers. There is a totally different framework of perception, food for them means something which fills the stomach and is a royal food.'"[39]

Prior to this intervention, the nutrition education provided in the hospital wards sought, first and foremost, to teach mothers that matooke and lumonde were not suitable weaning foods. Needless to say, as Dean and Jelliffe acknowledged in 1960, "Most mothers resent being told that they are giving their children the wrong food."[40] Moreover, because matooke was ideally at the center of every meal, every *kibanja* or garden, and nearly every important ritual or ceremony in Buganda, insisting that the green cooking banana was not a "good" food inevitably fell on deaf ears. As Stanfield explained:

> We taught them that matooke, which is a very poor protein food, is not a good food for babies, for children being weaned. And the mothers used to listen to us very politely, but they were obviously not really taking it in because matooke was a good food from their point of view. Food, from their cultural perspective, was something which filled the baby's tummy and satisfied hunger and it was also the food the kabakas had fed on for years.[41]

As expatriate physicians like Jelliffe and Stanfield became aware that this approach was inadequate and ineffective, they began investigating social and cultural facets of child health and nutrition in Uganda that could be encouraged as the basis of a public health approach to nutrition education.[42] This represented a shift away from earlier practices in which the presumed superiority and universal application of biomedical knowledge was diametrically opposed to "non-Western" institutions and philosophies. Instead, biomedical personnel slowly began to realize that effective health and nutrition education involved an awareness of local conceptions of food and illness—not all of which required rejection. They began to envision a public health model that represented a hybrid or synthesis of biomedicine and local cultures. Thus Stokes was able to convince unit personnel to "tell them that in fact *matooke* was a good food, but that it would have to be mixed."[43]

Combating protein malnutrition using mixtures of locally available sources of protein was, as we have seen, not a new practice in Uganda. The mandate of Dean's work in the early 1950s was to devise mixtures of vegetable proteins that could improve African nutritional health as he had done among malnourished children in postwar Germany. Although initially sidetracked by the immediate need for effective therapeutic measures, Dean's work remained focused on developing high-protein mixtures for commercial production—an endeavor that fit well with targeted, biomedical quick fixes, and the magic-bullet approach. The use of locally available mixtures of high-protein foods was also emphasized at Welbourn's child welfare clinics, both in the individual consultations where "the mother of a six-months old baby is advised about mixed feeding" and the group setting that very often involved cooking demonstrations.[44] One example cited by Welbourn encouraged mothers to feed their children beans through a poster depicting the process from cultivation and pounding to the preparation of a soup that could be poured through a sieve in order to make both a drink for infants and fried bean cakes for their older siblings.[45] The use of the sieve and cultivation methods involving "grass bunding to prevent soil erosion" also reflect the confidence in science and the "developmentalist approach" that prevailed in the 1950s.[46] Welbourn compiled a number of similar recipes that aimed to improve child nutrition in a Luganda booklet that was published in 1952 and subsequently translated and published in English under the patronizing title "How to Feed Your Child."[47]

Stokes first joined in a distinctly different effort, which sought to employ local rather than purely scientific concepts of food, in the early 1960s, prior to the opening of the unit and while she was still in charge of the Domestic Science course at Gayaza High School. Together with Jelliffe and his colleagues at Mulago, Stokes developed several recipes for what were called "*ettu* pastes," or mixtures appropriate for children as young as five months old containing matooke or lumonde together with both vegetable and animal-derived proteins. With the exception of the eggs, fish, and dried skimmed milk, the remaining ingredients of ettu pastes were identical to the basic components of the average Ganda meal and they were prepared by wrapping them in a special *luwombo* or banana leaf and then steaming them together in the same pot with food for other meals.[48] The advantage, as Stokes and Jelliffe argued, was that ettu paste was prepared "using the traditional method of cooking," and was therefore "economical of fuel, as it would be cooked along with the family matooke."[49] The Luganda concept ettu referred to the separate banana leaf or "special packet" of food that would, it was hoped, ensure that young children consumed sufficient quantities of protein along with the staples of the Ganda diet. Ettu pastes aimed to encourage mothers to prepare separate foods specifically for young children. Stokes and Jelliffe hoped that focusing exclusively on the need to modify a single component of food preparation—the separate packet or leaf—would encourage mothers to readily adopt and implement the practice in their own homes.

Before long they discovered that the disadvantage of the ettu concept was that it said nothing about the contents of the packet itself, and therefore had limited value as a means of nutrition education. What was needed was a local concept capable of capturing the importance of mixing different foods. Stokes and her team then hit upon the potential of *kitobero*, which Mike Church referred to as "a cultural breakthrough."[50] Kitobero was already "a popular type of food mixture including many types of food," and the theme of a popular song frequently played on local radio stations. The lyrics of the song encouraged the cultivation and consumption of local high-protein foods and emphasized the prosperity and well-being associated with "self-help" practices:

> Let's cultivate groundnuts,
> We will cultivate empindi (cowpeas),
> We will cultivate sim sim (sesame),
> Then we can buy good things,
> If our backs are strained we'll recover strength to bend and sow with our hands.
> Listen! Ekitobero. Then every year we'll have Ekitobero,
> Kalembe, you start cultivating! . . .
> Let's cultivate and we'll grow rich,
> My mother has cooked Ekitobero,
> Then every year we'll get Ekitobero,
> I have played this guitar on Ekitobero,
> Ah! Ah! Ah! Ah!, Hiii Ekitobero,
> Ha hahe![51]

The song endorsed "the food in very positive terms" and mothers often initially responded to kitobero with smiles and laughter—apparently echoing the musician's laughter midway through the song.

Mothers and other Ugandan visitors to the unit understood that kitobero contrasted with more conventional definitions of "food."[52] As was common in many African societies, Ugandans interpreted "food" or *emmere* as a reference to the staple component of the diet, which fills the stomach and alleviates hunger, and which in Buganda was commonly a reference to matooke. Although emmere was always served and eaten with enva, they were conceptually distinct and usually prepared separately. Related to the verb *tabula* meaning "to stir up or mix" and to conceptions of being mixed up, mingled, muddled, or confused, preparing kitobero entailed mixing matooke with the otherwise separate components of the meal as part of meal preparation and thus before eating, which was otherwise when matooke or another emmere was usually mixed with enva. In Luganda, the word *olutabu*, used to designate soft and bland food or "pap" suitable for infants, was, in fact, derived from the same verb, tabula, and, while kitobero was not a food solely eaten by young children in Uganda, it had the added benefit of being by definition an appropriate first food for babies and infants. This was another reason why the idea that matooke was a "bad" weaning food simply did not make sense to mothers in Buganda, as matooke was not only filling, easily satisfying a child's hunger, but was typically steamed and mashed into a paste that, like olutabu, was an ideal consistency for even very young children. Etymologically kitobero is also related to *e-kitobeko*, meaning "a mixture or variegated pattern" as in a multicolored mat. Moreover, the prefix *e-ki-* indicates that something "is good," so kitobero thereby simultaneously conveyed the notion of mixing otherwise separate components of the diet and the notion of a food that is "excellent" or as one acquaintance explained, "a feast."[53]

The rich and positive meanings embedded in the concept kitobero highlighted the act of mixing different foods and learning how to prepare and teach others to prepare kitobero, as a weaning food for young children became the centerpiece of what mothers learned at Mwanamugimu. Like ettu pastes, the recipes for kitobero consisted of either double mixtures of the most common staples, matooke or lumonde, together with vegetable proteins or triple mixtures that also integrated available sources of fish, meat, or eggs. In keeping with the stipulation that the proposed modifications in feeding practices be kept to an absolute minimum, women learned that preparing kitobero required diverging from customary methods of food preparation in only two respects. First, the ingredients had to be mixed before they were steamed and mashed, rather than as part of the consumption process, in which portions of the cooked staple were typically dipped into the relish or enva as they were eaten. It was in this respect that the concept of kitobero underlined the second innovation, which involved cooking the mixed food for young children in a separate banana leaf or, as became the common practice at Mwanamugimu following the request of mothers, in a small, lidded pot or "box" known as *akaboxi*.[54] Women and men who learned how to prepare kitobero learned that the mixture could be made from the foods available in their *bibanja* or gardens and local markets, as the precise combination of the ingredients was of little consequence. What did matter was that the dietary components were mixed in amounts that ensured the consumption of sufficient calories and protein. The key to making kitobero specifically for young children was, as many of my informants impressed upon me, to measure the ingredients. In fact, measuring to ensure a balanced mixture was one of the most enduring messages of the Mwanamugimu program. Women who instructed those new to or visiting the unit demonstrated how to measure two handfuls of soaked beans or pounded groundnuts and to add these ingredients to either three "fingers" of matooke or to a medium-sized tuber of lumonde. With the exception of the spoonfuls of pounded *nkejji*, or small dried cichlids commonly sold on skewers, the basic unit of measurement employed at Mwanamugimu was the handful (see fig. 3.7). When asked about Mwanamugimu, informants from the region surrounding the Luteete Health Center frequently responded by emphatically holding out their open palm to illustrate how to measure the high-protein components of the mixture, and very often they would proceed to teach me the various steps involved in its preparation. After laughing when I asked about kitobero, Maria Zerenah Namusoke explained how to put the ingredients "together in an akaboxi and then . . . cook it before feeding it to your children."[55] Their memory of measuring the ingredients reflects the emphasis that was placed on ensuring consumption of both calories and proteins, or "body-building" and "filling" foods, as they were known at Mwanamugimu.

Although I did interview a handful of women who claimed to not know how to prepare kitobero, the vast majority indicated that kitobero was a well-known dish often prepared for young children. Peragiya Nanziri's testimony exemplifies what I commonly heard in the field: "When I was still producing I used to cook kitobero for my children."[56] What is more, inquiries about kitobero inevitably led to discussions of its virtues: I was repeatedly told that "kitobero is a good food for babies and children . . . children [fed kitobero] grow well and their bodies are stable."[57] Fatuma Nankabirwa explained, for example, that kitobero "is a good food because it is a building

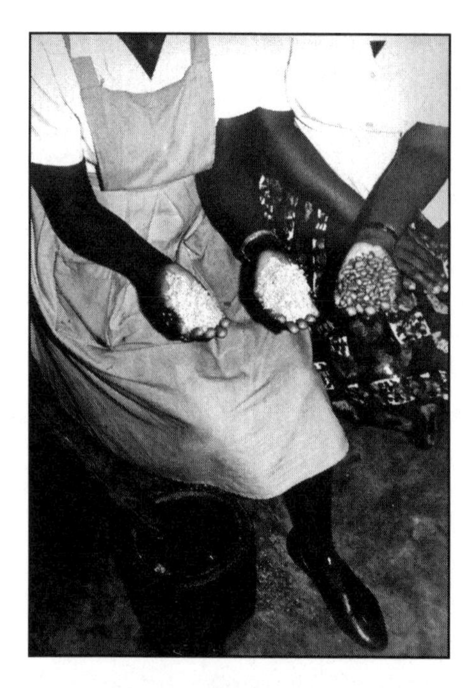

FIGURE 3.7 AND 3.8. Measuring using a handful and informal discussions. Courtesy of Paget Stanfield.

food."[58] Catherine Nansamba insisted that "the life of babies who take kitobero is different from those who don't take kitobero," and when I asked why, she responded that "they look good and healthy."[59] An especially elderly woman, Solom Nanyonga, even proclaimed, "Let them continue with kitobero as it was in the past because children grow healthy and powerful."[60] The extent to which kitobero and the practice of measuring and mixing ingredients became a part of the living memory of these women and men is a testament to the perceived efficacy of kitobero in the treatment and prevention of severe acute malnutrition. It is also a testament to the power of this synthesis of the local and the biomedical in public health promotion.

Beyond the "Lecture-Poster Barrier"

The importance of measuring and mixing the body-building and filling foods was a central theme of the daily cooking demonstrations as well as a number of innovative instructional techniques that became an integral part of gaining and imparting knowledge at Mwanamugimu. Recognizing the limited efficacy of conventional didactic approaches or what became known as "the lecture-poster barrier," the program rapidly abandoned standard educational techniques in favor of less formal means of conveying information.[61] Over time a high value came to be placed on informal discussions between mothers, which were seen as "group learning" opportunities that allowed "mothers [to] teach and influence each other. . . . The 'Group learning' that happens between mothers involves practical skills, encouragement and sympathy, as well as knowledge; the result is as dramatic a change in the mothers as in the children." Rather than direct or interrupt these invaluable discussions, the unit staff listened, answered questions, and provided essential feedback (see fig. 3.8).[62] Moreover, it was found that an added advantage of encouraging interaction between mothers as a part of the program itself was, as John Kakitahi, the later director of the unit, explained, that it lessened the anxiety of mothers who feared for their children's lives. According to Kakitahi, mothers at the unit were able to reassure and comfort their peers far better than biomedical personnel, and through their own positive experiences they were able to provide essential assurance and support.[63] Over time a variety of additional innovative pedagogical tools, including songs, stories, proverbs, plays, photographs, and calendars, came to augment the demonstrations and discussions that originally formed the centerpiece of the nutrition education at the unit.

The Mwanamugimu staff came upon the idea of publishing instructive calendars during the routine follow-up home visits conducted to both check on previously malnourished children and provide further health education to mothers and their wider community.[64] After observing that calendars hung on the walls of nearly every rural home, they created instructive calendars in order to further publicize the message of Mwanamugimu and the value of kitobero. In the 1969 calendar, for example, images and text proclaiming the virtues of kitobero and listing the ingredients and how to measure them adorned the months of March through November. Moreover, the photographs above June and August displayed handfuls of beans, groundnuts, meat, and fish below bold text that read *ebipimo*, or "measurements"[65] (see fig. 3.9).

As part of the central importance placed on the idea that seeing is believing, photographs also became a method of graphically illustrating to both participants in the program and those back home that severe acute malnutrition was a condition that could be treated and prevented through the

FIGURE 3.9. June 1969 calendar. Courtesy of Paget Stanfield.

provision of kitobero. Thus heart-wrenching photographs of severely malnourished children with their visibly anxious and distressed mothers and guardians were taken on arrival at Mwanamugimu. These images were then placed beside photographs taken immediately before the children returned home. The contrast between these two very different images was striking, as both mothers and guardians and their formerly malnourished children had, in a few short weeks, undergone a very visible transformation. In the "after" shots, the mothers and guardians were frequently smiling broadly, their satisfaction and relief readily apparent. The children themselves were often so changed that they were not immediately recognizable as the same children in the images taken only a few weeks before. Moreover, their misery and suffering had been replaced by the playful curiosity and joy common among children their age. Mothers and other women who were trained in the principles of the Mwanamugimu program received certificates of their newly gained knowledge and expertise. In addition to bearing the signatures of Stokes, Faith Lukwago, Jelliffe, and other pediatricians involved in the program, these official certificates included the before and after shots so that they could then be shown to relatives, friends, and acquaintances, thereby spreading the knowledge of Mwanamugimu and kitobero within their networks and communities (see figs. 3.10 and 3.11).

The power of song to influence and instruct also became a central component of the pedagogical philosophy at the Nutrition Rehabilitation Unit. In addition to spontaneous singing and dancing, mothers composed songs, and the value of measuring and mixing kitobero was the subject of a number of Mwanamugimu songs. A "Kitobero Song" served, for instance, as a chorus reciting throughout *A Play about Kitobero*: "to cook and give children *kitobero,* that all children need to grow is *kitobero.*"[66] Through short "work and slogan" songs, mothers punctuated key points and techniques in their cooking demonstrations, including a set of lyrics to accompany the process of scraping meat to make it an edible ingredient in kitobero prepared for very young children. Music emerged as a means of spreading nutrition education, complementing and reinforcing the messages of pictures and illustrations beyond the walls of the unit.

After several failed attempts to commission a song by writing lyrics for a musician to "translate and adapt," the staff invited a popular Ugandan musician to visit Mwanamugimu in order to compose a song. Dan Mugula was the musician chosen by the staff at Mwanamugimu because he was a very well-known *kadongo kamu* (ballad singer) heard frequently on Ugandan

MWANAMUGIMU CLINIC

Laba bwebatangalijja, laba bwebakayakana nobulamu obulungi!

Anti kiva mukulya emere eno. NUNGIWA

Amaze mu kisulo kya MWANAMUGIMU sabbiiti ...4..... nga ayigirizibwa ebyabaana: ENDIISA, ENZIJANJABA NEMBERA Y'OBULAMU, NEBYOKUKOLA NEMIKONO.

Mrs. C. N. STOKES — 7 July 1966
Mrs. K. KIBUKA — PROF. D.B. JELLIFFE
J. MAFIGIRI — Dr. I. Schneideman

FIGURES 3.10 AND 3.11. Mwanamugimu certificates. Courtesy of Paget Stanfield.

radio stations and was especially known for writing songs of political and social importance.[67] Mugula was granted complete artistic license and, as Church emphasized, "no text or suggested contents were given."[68] Within four months of his visit, Mugula recorded a song detailing how to prepare kitobero and treat and prevent malnutrition for radio station listeners throughout the region. As the following excerpt illustrates, a significant proportion of the five-minute song encourages and instructs mothers to cook kitobero for their children.

> When your child reaches the age of 6 months . . . what you have got to do is . . . mix its feed . . . . The right measurement is also very precious. . . . For the right lunch you have to prepare the right "kitobero." You've got to do this with groundnuts you measure one handful, and pound them. . . . Three fingers of matoke are just enough with beans . . . . Also here, dear friend, you measure a handful. . . . With all these, friend, you've got to measure with your palm. Cook now till it's ready, mash it and keep apart the child's supper, then give to the child. That's the right kitobero and you will really see the change there is! Even if the child has ceased growing or does not grow the child will start growing.[69]

Mothers also created and performed plays that implored visitors and peers to measure and mix food for their young children. *A Play about Kitobero* begins, for instance, with a "mother" overhearing several conversations praising kitobero as a food containing all that children "need to grow and be filled." The mother, clearly depicting a woman who had not participated in the Mwanamugimu program, then asks a "teacher," "What is this Kitobero that everyone is talking about? . . . . Now tell me how I can make a good Kitobero." Interspersed with several songs, the play is essentially a conversation in which the teacher describes how to make kitobero for the inquisitive mother.

> You need to measure enough for lunch and supper. . . . Take a handful of dry beans, soak and peel them and that makes 2 handful of soaked beans . . . now take a filling food—let's say *matoke*, 3 small fingers and cut them up. . . . You can wrap them in a luwombo, but if possible it is best to put them in a akaboxi. . . . And when it is cooked, mash it well and divide it.[70]

At the end of the play, the mother then replies, "That's wonderful. Thank you very much for your *magezi* [wisdom]—now I must quickly go home and start practicing to be a good builder of my children by mixing good Kitobero."[71]

This emphasis on promoting nutritional health through the knowledge and wisdom gained at Mwanamugimu was also evident in another play entitled *The Story of the Builder.* This play featured the building of a new house using concrete blocks. However, Mr. Musoke—the builder and main character—attempts to save money by mixing one-third of the recommended amount of cement, and after a few months the blocks are badly cracked and one of the walls is crumbling. A *fundi* or "expert" who happens to pass by exclaims, "Goodness gracious! It looks as if you have a weak cement mixture there. . . . Musoke you really should learn how to mix things properly first."[72] The play concludes when Musoke's wife, who arrives in time to overhear the end of the conversation between her husband and the expert, draws an analogy between her husband's crumbling building and her malnourished child:

> Do you know that's just what they were telling us at the clinic today. . . . I took our little Semwanga there because he was swelling and so miserable, even his skin was breaking down. . . . Other mothers at the clinic demonstrated how to measure building food. . . . They were just like *fundis* . . . because their children I saw were strong and fit and hair so black. . . . Now I see it clearly—a building needs the right mixture of sand and cement to be strong. And a child needs the right mixture of building food and filling food to grow and be strong.[73]

Over time parents and guardians of once-malnourished children were also given the ability to trace their child's rehabilitation through growth charts. These charts of a child's weight, height, and age were initially viewed as a medical record, providing essential data purely for biomedical personnel as they monitored growth in order to identify early signs of malnutrition. Employed as yet another tool of nutrition education, weight charts became an increasingly important component of the Mwanamugimu program by placing the ability to monitor and manage a child's growth in the hands of their mothers and guardians. Thus weight charts further equipped and empowered them in the management and promotion of nutritional health. As Church and Stanfield argued, "Contrary to the general assumption that mothers cannot comprehend weight graphs, in practice . . . they do appreciate the 'road of healthy growth' and the deviation of the 'road of kwashiorkor.'"[74] As such, teaching women to read and interpret growth charts became a feature of the health education provided at the unit, and bringing a child's growth chart to follow-up clinics and regular checkups became a practice that is still seen today (see fig. 3.12). In line with the unit's central philosophy that a healthy child comes from a healthy mother, weight charts eventually expanded to include the prenatal period of a child's growth by also charting the weight gain of expectant mothers in order to encourage health and nutrition during pregnancy. The program also began to encourage birth spacing through innovative teaching methods and agricultural analogies including garden demonstrations that illustrated how thinning seedlings to provide adequate room and nutrients for growth translated into stronger and healthier plants and thus stronger and healthier children.[75]

Additionally, weight charts provided opportunities to evaluate the nutrition rehabilitation program and its impact. Growth charts were capable of capturing the rehabilitation of individual children, and when examined in

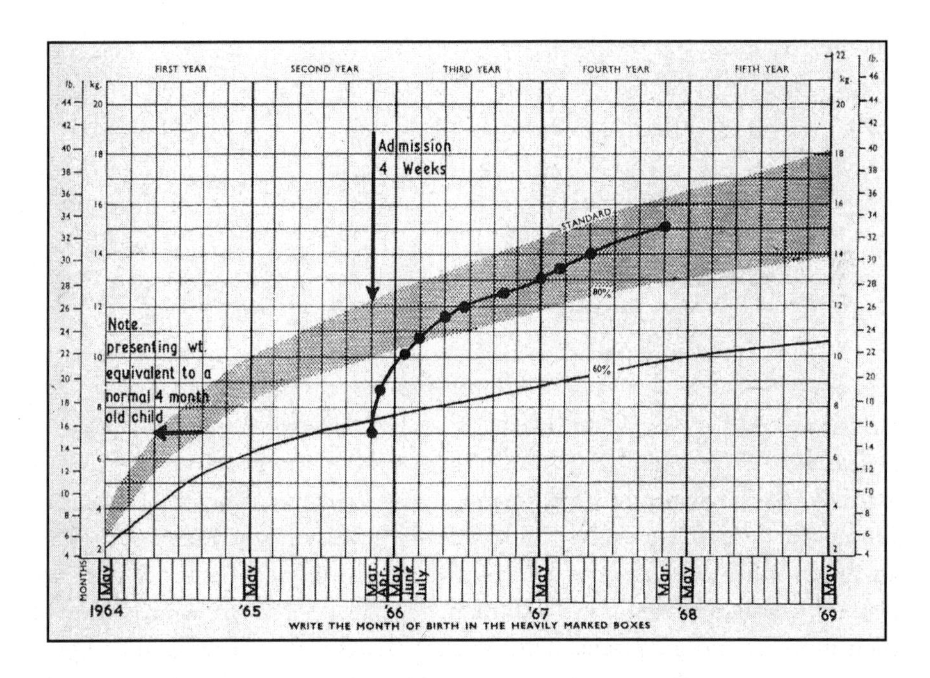

FIGURE 3.12. Weight chart. *Source:* Stanfield, "Malnutrition in Uganda," 52. Courtesy of Paget Stanfield.

aggregate and compared to evidence from siblings, they told a remarkable story of effective and sustainable prevention—notably after more than three decades of nutritional research and failed attempts to prevent malnutrition in Uganda. One such assessment found that in a group of seventy-three children attending the unit over a six-month period, 86 percent exhibited rates of growth that would allow them to recover the ground they had lost as a result of their experience with malnutrition, and 23 percent were gaining weight at rates that exceeded the average for their age. This study found that only ten of the children, or 14 percent, were not gaining at a rate sufficient to erase their weight deficit.[76] An analysis of over two hundred children who were either moderately or severely malnourished at their first visit to the unit, with an average weight of 63 percent of the standard weight for children their age, found that by their seventh visit to Mwanamugimu they were, on average, gaining weight at rates considered standard for their age. Thus, not only were these children no longer suffering from severe acute malnutrition and returning to the hospital for repeated treatment as cases of relapse, they were on a "road of healthy growth."[77] Growth records for siblings of children who had been brought to Mwanamugimu suffering from malnutrition were equally impressive, with the median growth curve of forty-five siblings in one analysis falling well within the range of healthy growth and development.[78]

The experiences of two severely malnourished children who were brought to Mulago in the mid-1960s further illustrate the impact that the program could have in young lives. One of the girls was almost two years old, but weighed less than sixteen pounds when she and her mother were admitted to the Mwanamugimu Nutrition Rehabilitation Unit. After only four weeks, the girl gained nearly two full pounds, and she continued to gain weight at such rapid rates that by her fourth visit to the unit—a short six months later—she had already caught up to other children her age. Over the next two years, as her weight chart clearly shows, she grew into a healthy, well-fed child (see figs. 3.12 and 3.13).[79] Another young girl, brought to Mulago by her pregnant mother, suffered from many of the classic symptoms of severe acute malnutrition: she was visibly miserable, swollen, and the skin on her legs was peeling away. Her mother's distress and anxiety is also palpable in the surviving grainy black and white image (see fig. 3.14). Before they left Mwanamugimu, a short three weeks later, the mother had given birth to her second child, and in the parting image taken before they returned home, she is seated together with her two healthy children, smiling—the earlier concern and fear replaced by a clear sense of pride and joy (see fig. 3.15). Images taken during follow-up visits to the unit over the next five years show the eldest child growing up together with her younger siblings, apparently free from malnutrition and exhibiting healthy rates of growth and development (see fig. 3.16).[80] These children and others like them provide living proof of the value of a hybrid public health approach to the problem of severe acute malnutrition in Uganda.

FIGURE 3.13. Mother with her daughter during a follow-up visit two years after admission. Courtesy of Paget Stanfield.

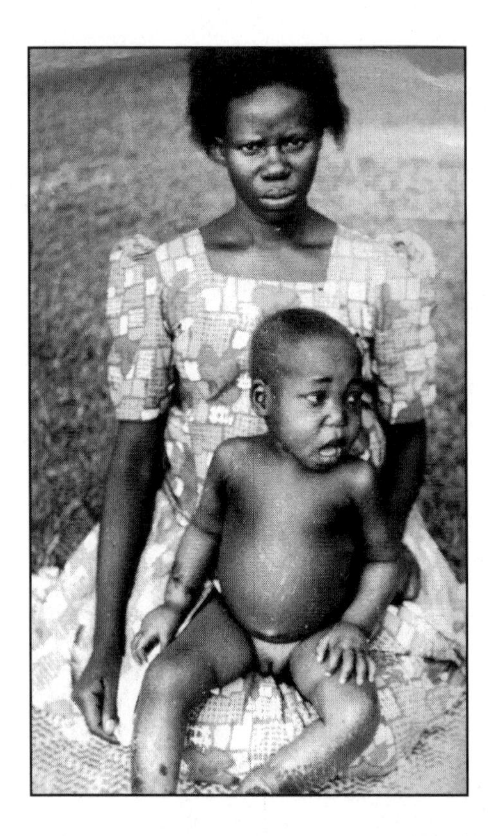

FIGURE 3.14. Mother with her severely malnourished daughter on admission. Courtesy of Paget Stanfield.

Influencers and Rural Outreach

In addition to the development of novel pedagogical methods, as the program evolved, emphasis was also placed on spreading the knowledge of kitobero to rural communities beyond the reach of the unit. From the outset Mwanamugimu was an important training site for medical students at Makerere who would later encounter severely malnourished children in their practice, but the emphasis of the training component of the program shifted to focus primarily on influential Ugandan women who might play a part in the promotion of nutritional health through their perceived roles as wives, mothers, and community leaders. At the close of an article describing Mwanamugimu for the journal *Mother and Child*, Church wrote, "Maybe when 'Kitobero' for children becomes the pattern in every home and children will

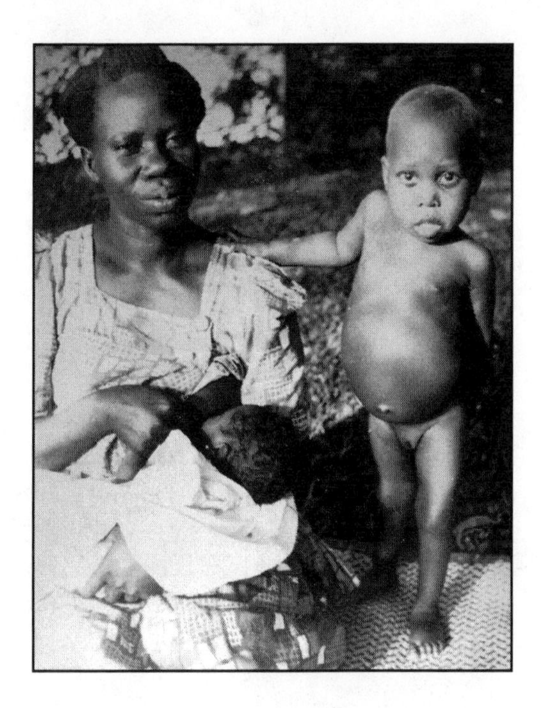

FIGURE 3.15. Mother with her daughter and second child after three weeks at the unit. Courtesy of Paget Stanfield.

FIGURE 3.16. Mother together with her subsequent children after five years. Courtesy of Paget Stanfield.

not inevitably tread the road of Kwashiorkor then the hope will be fulfilled." Further development of the Mwanamugimu program then focused, as Church explained, on "quietly" diffusing "the message . . . deep into the countryside where it is needed most."[81] In this period, the effort to extend the program into the countryside involved training individual women and, importantly, the first rural incarnation of the program at a health center in the village of Luteete.

The training of Ugandan women at Mwanamugimu initially focused on the mothers of formerly malnourished children, who were seen as the primary conduits of the Mwanamugimu message. After returning from the central unit to their homes, these mothers and guardians of once-malnourished children would, it was hoped, serve to endorse the philosophy of the program by providing an example and teaching what they had learned to their peers. However, it was soon realized that the mothers of formerly malnourished children were not the influential role models they were thought to be, and efforts were increasingly directed toward women who already served as community leaders and could more effectively persuade others to prepare kitobero for their children. Aptly known as "influencers," these women became a fundamental component of the nutrition rehabilitation program and its extension into the rural areas beyond the reach of the Mwanamugimu unit.[82] Nabanja Kaloli, for instance, trained to become an "influencer" at Mwanamugimu while pregnant with her first child. This training and her subsequent work teaching the principles of the program to young mothers in her community was especially important to her after her husband forbid her from working as a schoolteacher following their marriage. For Kaloli, training in the promotion of nutritional health at Mwanamugimu allowed her to gain knowledge and expertise that she could impart to others within her community, and was, therefore, a clear and multifaceted form of empowerment.

Women who underwent training at Mwanamugimu, either as future influencers or as mothers of once-malnourished children, then began to form and join Mwanamugimu clubs. In the postwar period, the Ministry of Social Development began to actively assist in the establishment of community clubs involving ordinary women from rural areas who were not part of the educated elite. According to Aili Tripp, "the initial focus of the clubs was on child survival because this was of concern to African women. . . . Few Ugandan women take for granted the importance of knowledge of nutrition, basic hygiene, and childcare given that this knowledge saves lives and prevents serious illness."[83] Moreover, these voluntary associations, despite their practical emphasis on domestic skills, paved the way for Ugandan women to take more active positions within the political sphere. For many women, community clubs and groups provided an opportunity to acquire new skills and become involved in initiatives within the community.[84] Building on the significant expansion in women's groups and clubs beginning in the late 1940s and early 1950s, Mwanamugimu clubs became yet another means of diffusing knowledge of the program to rural communities. The effort to train women at Mwanamugimu to become influencers explicitly drew on the growth of women's associations, and women involved in women's clubs and organizations were increasingly invited to spend two weeks at the Unit in Kampala learning the principles of the program and witnessing for themselves how mothers could treat and prevent malnutrition with kitobero.

In addition to what Mwanamugimu clubs and influential women like Kaloli may have accomplished, for one rural community the Mwanamugimu program was brought much closer to home. In the mid-1960s the newly independent Church of Uganda granted Stanfield and Church permission to take over the weekly clinics held at one of the three maternity and child welfare clinics that remained under church control. These maternity and child welfare clinics were originally built in the interwar period by the Church Missionary Society as part of an effort, led by Albert Cook and his wife, Catherine, to combat high levels of infant mortality in Uganda. In the years following the Great Depression, all but a selection of key centers were either closed or relinquished to government control. Luteete was kept open and became the chief maternity center, due in large part to support from the government of Buganda. As the center was built on land held by the Kabaka and was only a short distance from the Kabaka's rural palace in Bamunanika, and because officials within the Kabaka's government supported the maintenance and expansion of the center, Luteete continued and continues to this day to offer a range of maternal and child welfare services.[85] In the early 1960s, midwives stationed at the center provided care to new and expectant mothers and their young children, and a physician from the former CMS hospital at Mengo visited on a weekly basis.

Inspired by their involvement in Mwanamugimu and the principles of preventive social medicine, Stanfield and Church launched a comprehensive public health program at Luteete. Over the next several years and with a very small grant from the international charitable organization, Oxfam, the small maternity and child welfare clinic grew into a vibrant hub of activity centered around a community endeavor to promote health and well-being. Despite the many parallels with the state-of-the-art teaching facility established at the Kasangati Health Center, the initiatives at Luteete were also unique. Not only was Luteete far more rural and rudimentary, but, in line with the program at Mwanamugimu, far greater emphasis was placed on community involvement and oversight. The man recruited to direct the center, Luka Mukasa, was already a well-respected member of the local community and reportedly became "the pivot of all of the activities in Luteete."[86] Like Gladys Stokes, Mukasa was chosen for his leadership qualities rather than his prior medical training. What is more, a local *lukiiko* or council had ultimate authority over the center and significant influence over the initiatives implemented as part of Luteete's expansion.

One of the signature projects of the Luteete Health Center was the construction of a series of protected springs. The project itself was initiated by the lukiiko in charge of the health center, as providing a supply of safe water was deemed a "felt need by the local community."[87] Whereas water had previously been collected from stagnant pools, following the project it flowed in a clear stream from pipes connected to cement walls designed and built by a local mason. The involvement of the entire community made the construction of the protected springs particularly memorable; women I interviewed recalled how they collected the stones, while men remembered digging and pouring the cement. Moreover, the success of the first protected spring encouraged even greater participation in the construction of three additional protected springs, with the local pastor and *miruka* or subcounty chiefs contributing (see figs. 3.17 and 3.18).[88] Additional projects also reflected the emphasis on environmental sanitation and home improvement, including the construction of small pit latrines for very young children and the building of a model home with bricks made using an innovative brick-making device that could produce high-quality bricks from the local laterite soil and only a very small amount of cement. This model home was known as *bwamaka bulungi* or "beautiful home" and served as a rural Nutrition Rehabilitation Unit where local parents and guardians could bring malnourished children and thereby avoid the longer journey to Mulago.

Where the initiatives at Luteete diverged most explicitly from Mwanamugimu was in the emphasis placed on poverty alleviation and the recognition that the role of poverty in childhood malnutrition could no longer be ignored. Prior to the establishment of Mwanamugimu, the assumption was that economic factors were negligible and this was seen as especially true

FIGURES 3.17 AND 3.18. Protected springs pictured during construction and following completion, Luka Mukasa seated on right. Courtesy of Paget Stanfield.

when it came to severe acute malnutrition. In an application submitted to the Rockefeller Foundation for a pilot study of the Nutrition Rehabilitation Unit, it was stated quite clearly that "in Buganda, there is no doubt that the prime cause is not poverty or lack of protein food, but is the incorrect usage of protein foods already available in the peasant's own garden, the market and the rural shop."[89] Yet as an increasing number of immigrant laborers brought their malnourished children to Mwanamugimu, physicians and staff were forced to acknowledge that "poverty was often an underlying factor in the development of malnutrition, and thus generating wealth amongst the poor was a valid component of preventive medicine."[90] As a result, the staff at Mwanamugimu began to direct their attention to the needs of people living in squalid and overcrowded neighborhoods on the edge of the city. Beginning in 1967, they offered a weekly outpatient clinic emphasizing, in both Swahili and Luo, how best to promote nutritional health using available cash resources and food obtained from local markets rather than one's own bibanja or garden.[91]

At Luteete, income-generating activities became a central component of the health center early on. With assistance from Makerere's Department of Agriculture, the swamp behind the health center was cleared and transformed into a smallholder farm demonstrating "zero-grazing" cattle-keeping techniques, poultry and rabbit husbandry, and the cultivation of nutritious vegetables, including a green, leafy vegetable known locally as *nakati.* To evaluate and exhibit viable smallholder farming methods, a tenant farmer and his wife moved to the health center in 1969 and lived at the center for a full year, with approval from the lukiiko.[92] In addition to farming techniques featuring nutritious foods and especially those that would improve household access to dietary protein, the demonstration farm at Luteete also encouraged the cultivation of new cash crops like pineapples that could capitalize on the fledgling but highly lucrative tourist market.[93] Under the leadership of Mukasa, a core group of the residents in the area around Luteete also formed a Tusitukirewamu ("let us all rise and help each other") club (see fig. 3.19). Throughout the late 1960s, members of Tusitukirewamu worked on each other's farms, cooperated in the transportation and marketing of their produce, and became centrally involved in the Luteete Health Center and its expansion.

Not all aspects of Luteete's comprehensive program survived the tumultuous period of violence leading up to and immediately following Idi Amin's coup. In fact, significant components of Luteete's program failed to have an

FIGURE 3.19. *Tusitukirewamu* club. Courtesy of Paget Stanfield.

enduring impact and have subsequently faded in the memories of even those who were most involved. Although members of the Tusitukirewamu club provided the most vivid accounts of the work they did at Luteete in this period, there were elements of Luteete's expansion that even they could not recall, until photographs helped to jog their memories. Even with these images, most had little to say about key aspects of the program, including the farming and livestock demonstrations. When pressed, many remembered that the swamp behind the health center had been cleared, but with the exception of the green leafy vegetable *nakati*, the demonstration farm seems to have had very little impact in their lives and failed to leave a lasting impression. The zero-grazing method of cattle keeping was perhaps the most difficult to implement and, as a result, the most difficult to recall. After repeated questioning, Ephraim Musoke and Bumbakali Kyeyunne explained that the type of cattle needed for zero-grazing was prohibitively expensive.[94] Only Nabanja Kaloli, whose husband served as the chairman of the Tusitukirewamu club, remembered building a chicken coop based on the model provided at Luteete, and although Kyeyunne claimed that other members of the community built rabbit hutches and kept rabbits as a source of meat, he alone among those interviewed remembered adopting this practice in his own home.[95]

While the desperate political and economic circumstances of the 1970s and 1980s may account for the failure of projects that required a substantial initial investment, other components of the program were retained or lost based on how well they resonated with the ideas and aspirations of the local community. Nearly everyone I interviewed remembered the building known as bwamaka bulungi and they vividly recalled this particular structure despite the fact that it had been destroyed and only a few remnants of its foundation remain (see fig. 3.20). Local memories of bwamaka bulungi center around the bricks that were used to construct the building, as they were made using a novel brickmaking device known as a "Cinva Ram."[96] Many who visited and participated in the activities of the health center wrote their names into the bricks as they were pressed through the apparatus and elderly residents around Luteete indicated that their names had been literally etched on the walls of bwamaka bulungi.[97] Moreover, at least one informant indicated that her home had also been built using bricks made with the Cinva Ram.[98] In addition to the model home, "a simple frame shelter" was constructed consisting of a large roof with only partial walls on two sides. Opening onto the road and the entrance of the health center, this structure was designed to provide an inviting, shaded, protected, and well-ventilated area for community meetings, demonstrations, and weekly health clinics, and as a potential model of architecture suited to rural facilities in the tropics (see fig. 3.21).[99] Within the local community, this structure became known as the "unfinished building" and, as a result, was later, as my informants put it, "finished."[100] The brick walls that now enclose what was once an open space stand as structural indication of one of the reasons why aspects of the comprehensive public health program failed in Luteete: Unlike bricks, which remain a highly valuable building material in the region, the frame shelter failed to fit with local conceptions of an improved structure or building.

Key components of the comprehensive public health initiative did have a lasting impact and form part of the living memory in the area served by the Luteete Health Center. Several of the women with whom I spoke constructed raised kitchens in their own homes in order to reduce the risk of burns for young children, although they were notably not rebuilt following the Bush War.[101] Digging a narrow and shallow pit latrine that could be safely used by very young children appeared to be even more widespread

FIGURES 3.20 AND 3.21. *Bwamaka bulungi* and the "simple frame shelter," Luteete Health Center in background. Courtesy of Paget Stanfield.

and, as a measure that improved sanitation and safety, many of the elderly women whom I interviewed reported digging small pit latrines when their children were still young.[102] Among the most memorable of the projects introduced at Luteete was the construction of the protected springs. In addition to the widespread participation in their construction, these simple improvements in water provision became part of the living memory and social practice within the community surrounding Luteete because, on a daily basis, people collected and continue to this day to collect a more sanitary supply of water from springs that they helped construct many years ago.[103]

Like the protected springs, Mwanamugimu and the preparation of kitobero have had an enduring impact on daily life and social practice in Luteete. The women and men whom I interviewed spoke extensively about their ability to prevent malnutrition as a result of their knowledge of kitobero and its preparation. According to Kasifa Kyeyunne, Mwanamugimu was responsible for "big changes in this area and that is why you see that there are no longer children suffering from malnutrition."[104] Robinah Kayemba, who was only peripherally involved in the other public health initiatives at Luteete, discussed the importance of preparing kitobero and her memories of the Mwanamugimu program at great length.[105] And Robinah Nanteza proudly recounted how she prepared kitobero for eleven years and never once had a child suffer from malnutrition and this, as she pointed out, was despite the fact that she had to work on neighboring bibanja in order pay school fees following her husband's death.[106] The majority of the women whom I interviewed in and around Luteete reported that they prepared kitobero for their own children, and subsequent visits to the region suggest that their testimony can be considered representative. It would appear that many of the children who grow up in the region around the Luteete Health Center have been fed kitobero as a matter of course. With only a few exceptions, the elderly women in the area around Luteete incorporated kitobero into their daily meal preparation as a weaning food intended to maintain nutritional health. Kitobero became a standard diet of their young children and a feature of their daily meal preparation in order to prevent malnutrition. In fact, only two of the women interviewed for this study reported preparing kitobero as a curative measure for an already malnourished child.[107]

Moreover, people in Luteete fervently believed in Mwanamugimu and the power of kitobero to promote the health and welfare of their children and grandchildren. If the certificates that they received after completing

Mwanamugimu training were not lost during the Bush War, they were kept as valued possessions. Nabanja Kaloli, for instance, had her certificate framed and hanging with her family photographs on her living room wall. Women who underwent training at Mwanamugimu, either as future influencers or as the mothers, grandmothers, or guardians of once-malnourished children, are the ones who formed and joined Mwanamugimu clubs. The extent to which the preparation of kitobero has become a widespread practice in and around Luteete is a testament to the ongoing work of these influential women. In the area around Luteete, members of a Mwanamugimu club participated in kitobero demonstrations at the health center every Wednesday for decades. With transportation often organized and provided by the local miruka chief, these women also traveled around to neighboring regions in order spread knowledge of kitobero preparation even further afield (see fig. 3.22). Robinah Nanteza, for example, first learned how to prepare kitobero while attending a demonstration organized by the local Mwanamugimu Club in a village a short distance from Luteete.[108] One of the elderly women I interviewed had been an active member of a Mwanamugimu club in Luteete until 1994 and shared pictures of herself with her fellow club members during our conversation.[109]

What is more, Florence Mukasa, who worked at the health center as a nurse-midwife, became so dedicated to Mwanamugimu that she made it her life's work to not only deliver babies and provide basic medical care to expectant mothers and young children, but to also prevent severe acute malnutrition in the community she served.[110] The people living in the region around the Luteete Health Center spoke at length about the late Florence Mukasa, and, recognizing her deep commitment to the program, they remember Florence *as* Mwanamugimu and continue to know her by this nickname to this day. Many of the women I interviewed had given birth at the maternity center under her guidance and care, and for many she was simply the midwife who "was good at delivering babies," and who had lived and worked as a central figure in their community, providing medical and midwifery services for decades.[111] For women like Peragiya Nanziri, Florence Mukasa was a midwife whose medical knowledge and expertise helped her to survive a difficult delivery: "[Florence] is why I am still alive today."[112] For others, both young and old, she was also a teacher who taught "them how to feed babies" and helped them to thereby ensure their children and grandchildren never faced the specter of malnutrition.[113] She even taught the principles of the Mwanamugimu nutrition rehabilitation program to the

FIGURE 3.22. Demonstrating how to prepare *kitobero* in rural area. Courtesy of Paget Stanfield.

now-elderly community health worker, Stephen Mulindwa Maseruka, who, as explored in the next chapter, continued to provide immunizations and other services at the Luteete Health Center in 2012.[114]

Although most of the women I interviewed are now grandmothers, they have passed their knowledge of how to prepare kitobero on to their daughters, sons, and their sons' wives so that they can prevent malnutrition in future generations.[115] In fact, Kaloli's daughter, Caroline, reported that she had already instructed her own daughters and nieces in the preparation of kitobero, indicating that Mwanamugimu has become part of the living memory and social practice in the new millennium and across several generations. As a result of this instruction from mother to daughter and son and from daughter to granddaughter, even very young mothers in the region have knowledge of how to prepare kitobero and plan to feed their children this mixture in order to promote nutritional health during the challenging period of weaning.[116] There are several factors that have played a part in the implementation of this minor modification in infant feeding in this one region of Uganda, including the efficacy of kitobero and the ease of its preparation.

Moreover, the Mwanamugimu Unit in Kampala continues to instruct women in the principles of preparing kitobero and, together with influencers like Nabanja Kaloli and Florence Mukasa, has kept the message of Mwanamugimu alive. Thus even in the context of the devastating period of political crisis and war that followed, Mwanamugimu came to be part of the living memory and social practice of both young and old in the area surrounding the Luteete Health Center.

By the late 1960s and after decades of nutritional work in Uganda, efforts to prevent severe acute malnutrition finally began making headway. Yet, the Mwanamugimu program was not simply the culmination of years of biomedical research and programming. It represented an initiative that emerged from a critical moment of reflection on the failure of prior endeavors. Effective curative therapies and the reliance on targeted biomedical solutions left physicians facing an increasing problem of relapse. Nothing could indicate the absolute failure of prevention more starkly than children returning to the hospital again and again, suffering from repeated bouts of a highly preventable condition. What years of research had achieved was the medicalization of malnutrition in the eyes of those tasked to devise a targeted weaning food and of those who brought their children to the hospital from ever greater distances seeking treatment. Biomedical practitioners may have thought that their curative measures created a passive reliance on the part of Ugandan women, but what they actually encountered was a local engagement with and interpretation of the therapies that saved their children's lives.

After facing the failure of their attempts to prevent severe acute malnutrition, a team of biomedical personnel based in Uganda returned to the drawing board. Inspired by pioneering programs elsewhere, they designed a program that sought to harness this local engagement with biomedical treatment and care, a program that sought to use the rehabilitation of individual children as health education that would, it was hoped, have an impact within the broader community. The resulting effort to demedicalize malnutrition and empower Ugandan women in the management of nutritional health led to the establishment of a hybrid local public health model known as the Mwanamugimu Nutrition Rehabilitation Program. What they endeavored to establish was a shell or framework into which a program could be built, a program with the flexibility to evolve according to the needs of the Ugandan people. Under the leadership of a locally appointed staff, the program developed into an innovative endeavor to prevent severe acute malnutrition by training mothers in the preparation and provision of kitobero. At Mwanamugimu, women learned that severe acute malnutrition could be treated and thus prevented with mixtures of food that they prepared themselves using ingredients and recipes that were already part of their daily cuisine. Through the work of mothers and influential women trained at the central unit and through the Mwanamugimu clubs that they formed, the principles of the nutrition rehabilitation program became part of the living memory and social practice within the region surrounding the Luteete Health Center—the first rural satellite of the Mwanamugimu program. According to the people in the region surrounding the Luteete Health Center, the children who are fed kitobero are healthy, strong, and no longer suffer from severe acute malnutrition. As a result of their faith in the principles of the Mwanamugimu program, knowledge of how to prepare kitobero has been passed down through several generations. In a final testament to the value of a program designed as a local public health endeavor, the devastating political crisis that followed undermined health systems in Uganda, but failed to fully disrupt the remarkable legacy and influence of the Mwanamugimu program.

4 IN THE SHADOWS OF STRUCTURAL ADJUSTMENT AND HIV

When I first visited the Luteete Health Center in 2003, I was completely unaware of later efforts to expand the Mwanamugimu nutrition rehabilitation program into a national program of prevention. I did know that Luteete had been an epicenter of the violence perpetrated by Uganda's first prime minister in both the 1960s and the 1980s, and I knew that under the notorious dictator Idi Amin, more than three decades of nutritional work had come to a decisive end. Knowledge of this devastating period of postcolonial crisis significantly limited what I expected to find in Luteete and fundamentally determined how I initially interpreted what I saw and heard. I arrived in Luteete unwittingly beholden to a framework of perception that read violent conflict and political crisis into the cracks of a crumbling infrastructure. I did not expect to find, and thus could not at first see, evidence of an innovative public health initiative launched more than thirty years before. Unlike the central Mwanamugimu Unit on Mulago Hill, which continued to admit malnourished children and their guardians in the first decade of the twenty-first century, little concrete evidence of Luteete's earlier expansion remained intact. The health center had since reverted back to a maternity and child welfare clinic and, even during my most recent visit in 2012, exhibited levels of disrepair and a shortage of supplies and personnel typical of rural health centers in Uganda and other parts of the continent. Bwamaka bulungi, constructed as a model home and rural nutrition rehabilitation unit, had been destroyed and never rebuilt, and the "unfinished" building remained, in 2003, dark and empty—a stark reminder of the time when, as Ephraim Musoke put it, "the health center was a powerful place."[1]

While Luteete's physical deterioration appeared at first glance to reflect the region's turbulent past, when I began to scratch the surface a very different picture emerged. What the current state of the health center obscures is the extent to which the nutrition rehabilitation program became part of the living memory and social practice in the area surrounding Luteete. In the midst of violent upheaval and waning donor support, the program survived, sustained by Ugandan women and men who worked to promote nutritional health in their communities and across generations. Through their efforts, the nutrition rehabilitation program became a truly local and sustainable public health initiative, dependent on local adoption and advocacy more than on national institutions and international donor funds. The remarkable survival of Mwanamugimu, therefore, speaks to the value that the nutrition rehabilitation program came to have in the lives of the women and men who have kept the Mwanamugimu message alive.

It also, as I came to realize, speaks to the stark reality of public health programming in late twentieth- and early twenty-first-century Africa. What I thought was solely the result of a devastating period of political violence and warfare turned out to also have been caused by neoliberal economic reforms and the resulting war on health. Where prior investments in infrastructure and expertise had kept the program alive through the Amin years, only people touched by the program could keep it afloat in rural regions against the tide of severe resource constraints and global health faddism.[2] The implementation of economic reforms, decentralization, and privatization in the late 1980s and early 1990s converged with mounting global concern about the HIV epidemic to inadvertently undermine support for Mwanamugimu. It is in the shadows of structural adjustment and global health that Ugandan people have promoted and continue to promote nutritional health in future generations. Their remarkable tenacity makes all the more poignant the extent to which their need to fill the resulting gap themselves is indicative of the withdrawal of both national and global interest and investment in nutritional health and programming in rural parts of Africa. Here then is the resilience that must surface when regions are largely abandoned by both a state shrinking under structural adjustment and nongovernmental organizations interested in other things.[3] In Luteete, the people have stepped in and, empowered by the Mwanamugimu program, worked to secure a healthy future for their children and grandchildren. It is a shameful reflection on global health that they have had to do so in the context of diminishing local sovereignty over health provision and an overarching divestment in integrated systems of medical care. It is also a sad reflection on global health studies that programs like Mwanamugimu remain largely forgotten outside of Uganda. The efficacy and staying power of the Mwanamugimu program illustrates that there is a great deal to learn from historical studies of prior efforts to contend with contemporary health challenges, and that even in the midst of especially dire circumstances it is possible to build health systems and public health programs that endure.

Asserting Sovereignty in the Midst of Crisis

Africa's first nutrition rehabilitation program fundamentally transformed the treatment and prevention of severe acute malnutrition in central Uganda. At first, very little changed in the realm of research. International experts funded by foreign institutions continued to direct the scientific study of malnutrition on Mulago Hill as they had been for decades. Yet, the focus of nutritional research in Uganda began to shift as the first decade of independence drew to a close and Ugandan experts and officials asserted their authority over the future of nutritional health in Uganda. Uganda was, for a brief moment, poised to continue at the forefront of nutritional science within a comprehensive national framework privileging the interests of the Ugandan people. However, the ambitious plans forged in this period never saw the light of day. Nearly forty years of nutritional research came to an abrupt end when Idi Amin seized power and unleashed a reign of terror that had devastating consequences. The unfulfilled promise of this immediate postcolonial period assumes even greater significance in the context of ongoing affronts to state sovereignty over public health in Uganda and other regions of Africa and the world.

In the late 1960s, nutritional research in Uganda entered a new phase, and this shift in the philosophy and approach to the science of nutrition is reflected in both the influence of the Mwanamugimu program and the changing political circumstances of the times. A 1967 UNICEF sponsored seminar on food and nutrition in Uganda, for instance, brought a diverse body of experts to the Ugandan capital, emphasizing the value of the interdisciplinary approach at the heart of this new chapter in the history of nutritional work on Mulago Hill.[4] The following year the MRC Unit came under the leadership of Roger Whitehead and Paget Stanfield. Whitehead had long been the senior biochemist at the MRC under Dean. Stanfield, as we have seen, was a key figure behind both Mwanamugimu and its extension to the Luteete Health Center. Although Whitehead and other researchers at the MRC Unit had at first been critical of the Mwanamugimu program, that skepticism eventually gave way, and for Whitehead at least, the public health philosophy of the program came to have a tremendous impact on his own thinking and approach.[5] As codirectors of the unit, Whitehead and Stanfield launched a program of research redefining nutrition as a science and practice of health promotion—a move explicitly influenced by the hybrid model of public health that Mwanamugimu represented. They renamed the MRC facility the Child Nutrition Unit, dropping both "infantile" and "malnutrition" and, at the opening of the unit's extension in 1969, publically announced the inauguration of a longitudinal study of nutritional health among children living in rural villages. Moving away from an emphasis on single rather than multifactorial theories of causation and a laboratory-based philosophy and practice, this "broad-spectrum approach" aimed to document the environmental and social factors influencing nutritional health and, in particular, the role of intercurrent infections, or what Whitehead referred to as "the nutrition-infection complex."[6] Blending "pure" and "applied" scientific work was equally fundamental to this new research agenda, and, as Whitehead later reflected, "this broader approach towards the aetiology of kwashiorkor and marasmus expressed itself in an increased emphasis on topics such as hygienic latrines, improved water supplies, water conservation and insect protection."[7]

The deputy minister of health, S. W. Uringi, also highlighted a more comprehensive vision in his speech at the opening of the MRC Unit's extension, where he notably stressed that "the Government of Uganda" was "making every effort to expand and extend medical services"[8] (see fig. 4.1). Key to this state-driven broadening of medical provision was greater coordination and direction from the Ugandan Ministry of Health and the National Research Council. This assertion of national authority over nutritional work in the early 1970s was, moreover, an extension of much earlier efforts on the part of Ugandan doctors to achieve recognition of their professional status and to decolonize medical practice and care.[9] Often coming under calls for Africanization, or the transfer of key posts to qualified Ugandans, Ugandan physicians worked to improve medical provision and in particular sought to substantially increase the number of doctors providing health care throughout the country. And these efforts bore fruit: in the early 1960s, the number of medical students at Makerere was set to double, and by 1969 the number of Ugandan physicians had increased more than twofold.[10] In that same period, Ugandan doctors assumed all administrative positions within the Ministry of Health, the Uganda Medical Association, and Uganda's medical board.[11] Thus significant progress had already been made in many areas of medical research and practice by the early 1970s, when the Ministry of Health assumed control of Mwanamugimu and the National Research Council called for the integration of the nutrition rehabilitation program,

FIGURE 4.1. Deputy Minister of Health, Mr. S. W. Uringi, delivering a speech at the opening ceremony of the MRC Child Nutrition Unit's Expansion in 1969. Roger Whitehead is seated on Mr. Uringi's right. *Source:* Official Opening of Unit's Extension, July 17, 1969, FD 12/281, PRO. Reprinted by permission of The National Archives.

the MRC Child Nutrition Unit, and the Ugandan Food and Nutrition Council. [12]

The National Research Council chairman, Professor John Kibukamusoke, was a highly successful researcher in his own right and a particularly prominent member of the Ugandan medical profession—also serving on Uganda's medical board and as the first African president of the Association of Physicians of East Africa.[13] Kibukamusoke was critical of the "ad hoc basis" on which nutritional research in Uganda had been conducted, and called for the creation of a national institute that, under the authority of the National Research Council, could ensure that the future program of nutritional research followed a locally determined set of priorities serving the needs of the Ugandan people.[14] Kibukamusoke's vision soon became an ambitious plan to construct a four-story, state-of-the-art building to house Uganda's National Institute of Human Nutrition. However, the MRC made it abundantly clear that it was not in the business of providing "aid" or subsidies, and Kibukamusoke's request that Whitehead serve as the scientific director of the institute until a qualified successor could be trained was rejected out of hand.[15] Instead, and admittedly acting more "as a longtime resident of Uganda" than "as an agent of the MRC," Whitehead served on the subcommittee appointed in April 1971 to devise "a national plan for Uganda."[16]

The National Institute of Human Nutrition, which was set to be established on Mulago Hill in 1973, represented far more than a simple amalgamation of its predecessors. The proposal involved the creation of two new departments and professorships in nutritional science and applied nutrition, which, through close affiliation with the Makerere University Medical School, would seek to integrate training with scientific research and applied programming. Moreover, the architectural plans and models of the four-story institute building reflect the importance placed on integrating numerous aspects of understanding and promoting nutritional health. Laboratories for fundamental research and the offices of professional scientists were to be adjacent to seminar rooms and facilities for experts and fieldworkers engaged in training, surveys, statistical analysis, community improvement projects, and school feeding programs. To facilitate cross-fertilization between different disciplines, the institute was also slated to employ experts in sociology, education, mass media, and rural extension and community development, as well as animal, game, and fisheries.[17] As Stanfield wrote, those involved in nutrition work in Uganda in this period realized they needed "to explore the boundaries (which are so often a no man's land) between medicine and such disciplines as sociology, educational psychology, agriculture, home economics, commerce, demography etc. The doctor is realising that he holds only a very small part of the total answer to malnutrition. . . . Without losing his identity he must develop links with these other disciplines who are providing the remaining parts of the total answer."[18] Whitehead later called this emphasis on interdisciplinary perspectives, multifactorial approaches, and an outright rejection of the division between fundamental and applied nutritional knowledge and practice the "'Uganda school' of nutritional thought."[19] The influence of the demedicalization of malnutrition at Mwanamugimu and the value of creating a space where such cross-fertilization would be able to emerge could not have been more clear. What is more, it was a school of thought that many believed held great potential for nutritional science within a comprehensive national framework of health promotion.

The political crises that engulfed Uganda in the immediate postcolonial period marked the end of this promising moment in public health programming. The government that assumed power when Uganda achieved independence was undermined by a number of unresolved political divisions. The privileged economic and political position that the Buganda Kingdom had enjoyed throughout the colonial period created deep-seated animosity between Ganda and many of the other ethnic and political groups. A power-sharing agreement between Milton Obote, Uganda's first prime minister, and the kabaka of Buganda, acting as Uganda's first president, did not last. The controversy over territory that inaugurated the first period of postcolonial violence in Uganda led to riots, which Obote responded to with military force. The physician and former nutrition researcher Eria Muwazi, who had become an important politician and Buganda's minister of health and works, was centrally involved and lost his position as a result. In the wake of further scandals, Obote rewrote the constitution and greatly enhanced his own authority. When faced with opposition to these affronts to democracy, Obote declared a state of emergency and ordered the Ugandan army to attack the kabaka's palaces, including the kabaka's properties in Bamunanika. With the help of a car and driver sent by Muwazi, the kabaka managed to escape over a palace wall and into exile.[20]

Word of the imminent attack reached Bamunanika and those living in nearby villages, including Luteete, and they immediately took measures to thwart the soldiers' advance. Men organized by a local chief dug a deep ditch in the road, hoping that when the soldiers found the road impassable, they would be forced to turn back. These men and others in the region were later arrested and imprisoned for periods reportedly ranging from several months to seven years.[21] When the soldiers did finally reach Bamunanika they not only ransacked the kabaka's palace but violently harassed the people living in the surrounding area, forcing many to flee for months. Food and property were confiscated, the midwives at the Luteete Health Center were raped, and women were sexually violated in front of their husbands.[22] Obote then abolished Buganda and the other Ugandan kingdoms, converting Buganda's parliament building and the kabaka's palaces into the new army headquarters and barracks.[23] Due to the proximity of the Luteete Health Center to the kabaka's palace at Bamunanika, those served by the first rural extension of the Mwanamugimu program also simultaneously suffered through a period of pronounced violence and upheaval. It is very possible that had Obote remained in power, the nutrition rehabilitation program may have never gotten off the ground in Luteete. As it turned out, Obote was overthrown in a military coup by Idi Amin, and the political repression and violence of Obote's first period in power (Obote I) meant that Amin's coup was initially celebrated by many in Uganda.

The ongoing violence of military rule did not immediately temper the euphoric mood, as Amin made numerous promises of reform and his brutal tactics of disappearance initially targeted only Obote loyalists.[24] Then in 1972, with the expulsion of an estimated seventy thousand Asian civilians and the first wave of civilian casualties, the "honeymoon period" came to a definitive end.[25] Over the course of his eight years in power, Amin turned to violence in order to shore up his waning hold on the state. Even before the ratcheting up of massacres and targeted killings, Amin's shifting international alliances and his decision to sever ties with Israel soured relations with Britain and other Western donors. The British High Commission responded by suspending further aid and instructing all British nationals to gradually and inconspicuously return to England.[26] In 1973, Stanfield and Whitehead were among the very last of the expatriate medical personnel to flee. Prior to their departure, they made final arrangements for Mwanamugimu and the MRC Unit to continue. The MRC ward that was previously occupied by severely malnourished children requiring intensive therapy became a small factory where the reinforced milk formula needed for treatment was produced and distributed to Ugandan hospitals. The treatment of these severely malnourished children was moved and integrated into the neighboring facilities at Mwanamugimu. Each week, Whitehead called the British High Commission and, assuming that all communications were being monitored, simply asked "Is there any change?" to which the reply was always "There is no change," meaning that another member of the expatriate staff had to then quietly leave the country.[27]

Amin specifically targeted educated elites, with Ugandan doctors at particular risk. George Ebine was the first physician who lost his life, reportedly for providing refuge to a pro-Obote officer. Vincent Emiru, Professor of Ophthalmology, was also killed in the first year of Amin's rule.[28] After the expulsion of the Asian civilians, an increasing number of medical personnel were massacred, and in response many fled the country. In March 1973, Kibukamusoke learned that his name had been placed on a list for elimination, and George Sembeguya, a prominent doctor and politician, was killed after being dragged from his surgery with a rope around his neck.[29] Although more than one hundred physicians had fled by the middle of 1977, many remained for as long as they could. The dean of the medical school, Alexander Odonga, only fled across the border to Kenya after receiving advance warning that he too had been singled out to be killed.[30] The highly renowned surgeon Sebastian Kyalwazi continued working at Mulago reportedly even after being listed as a target.[31] Those who remained in Uganda struggled to maintain medical services and told harrowing accounts of the obstacles they faced. According to Chris Ndugwa, the constant state of emergency involved long hours spent at the hospital working under a persistent fear for their lives. It was not safe for anyone to leave the hospital after dark, so hospital staff worked through the night, only to face a wave of patients arriving in desperate need of medical care early every morning.[32] Physicians not only feared for their lives but lived with the constant trauma of treating people who had been tortured and performing postmortem examinations on the corpses of those executed. They worked for rapidly diminishing salaries in facilities that suffered damage and lacked water, sanitation, refrigeration, and crucial medicines and supplies.[33]

By the middle of the 1970s, an era of nutritional research and programming in Uganda had come to a definitive close. The National Institute of Human Nutrition would never be built, and while Ugandans experienced an unprecedented escalation in violence, international attention shifted away from protein deficiency to focus instead on the infant formula controversy. Mwanamugimu began to recede into the long shadows of Amin's rule, and the innovative program of prevention was largely forgotten outside of Uganda. Few would remember that the East African country was once at the center of international efforts to promote nutritional health. And even fewer knew that the Mwanamugimu program continued against incredible odds to help a growing number of children recover from and avoid severe acute malnutrition.

Extending Services Even as Things Fall Apart

One of the key figures behind the nutritional work accomplished in Uganda in the 1970s and 1980s was Dr. John Kakitahi. Like so many doctors in Uganda, Kakitahi developed a passion for prevention after experiencing the frustration of treating children repeatedly for the same condition.[34] Like Namboze, Kakitahi left pediatrics very early in his career to pursue a postgraduate degree at Makerere's Institute of Public Health. In 1973, as Ugandans faced increasing repression and violence under Amin, he was recruited as a lecturer and then took charge of the MRC Unit following Whitehead's departure. Unaware of the existing research program or how to acquire the reagents needed for basic biochemical tests, Kakitahi struggled to keep the unit operational. The MRC buildings suffered little damage during this period but dwindling medical supplies combined with the difficulty of repairing equipment and procuring replacement parts ultimately made it impossible for the unit to continue to function as a research facility. Kakitahi was able to secure support from Whitehead and the Dunn Nutrition Unit in Cambridge (which Whitehead then directed) in order to go abroad to pursue further training in nutrition, thereby briefly escaping the escalating violence. At the end of his program in 1976, as Kakitahi later recounted, he "plucked some courage and came back" to Uganda and was immediately put in charge of the Mwanamugimu Unit.[35] Gladys Stokes, together with Faith Lukwago and other members of her team, had been able to keep the unit open, but as Kakitahi explained, it wasn't easy: "it was hell . . . things were rough . . . we had very few staff and . . . hardly any resources."[36] Yet, Kakitahi not only managed to keep the unit functioning by using his personal vehicle and what he could spare from his increasingly meager salary, but, together with Latimer Musoke, who had taken over as the Chair of Pediatrics following Stanfield's departure, also launched a program to train a cadre of public health workers and further extend the preventive promise of the Mwanamugimu program into rural areas beyond the capital city.[37]

It is difficult to make sense of an expanding public health program during Amin's brutal period in power. At first glance it appears to counter the prevailing narratives of devastation in nearly all aspects of Ugandan life. What's more, we know that for medical personnel, Amin's Uganda was an especially dark time, and for the infrastructure so essential to health provision it was a period marked by deterioration and decline.[38] There can be no question that the circumstances in late 1970s Uganda were extremely dire. Kakitahi's return from Cambridge coincided with the dramatic intensification of violence that followed an Israeli attack on Uganda's main airport at Entebbe. In June 1976, a plane departing Tel Aviv for Paris was hijacked and with Amin's blessing rerouted to the Entebbe airstrip. A week later Israeli forces rescued the hostages in a predawn raid that left the hijackers and twenty of Amin's men dead. The Entebbe air raid represented a substantial defeat for an increasingly beleaguered military regime and Amin responded by terrorizing the people of Uganda.[39] Precise figures may never be known, but it is clear that the already considerable number of people murdered by state security agents surged in the final years of Amin's rule, and an estimated ten thousand people disappeared from mid-1976 through late 1977 alone.[40]

In Amin's Uganda, Mwanamugimu became an opportunity to promote health and welfare at a time when the provision of medical care narrowed. The remarkable sustainability of the Mwanamugimu program brings into focus what is often lost in the shadow of shallow tropes depicting Africa as a continent defined by crisis. Mwanamugimu's expansion during what may have been Uganda's darkest hour testifies to the crucial value of long-term investments in comprehensive public health programming *and* biomedical expertise. Both Musoke and Kakitahi were products of East Africa's finest institutions of higher learning and medical training, and they were able to marshal their knowledge and skills to make the most of the unstable and insecure realities of their time. Kakitahi was part of a generation of medical students trained specifically to increase the number of physicians serving the Ugandan people. The basic infrastructure, simple technology, and flexible framework of Mwanamugimu made it a local program capable of continuing in the context of shrinking resources. And while this is something to celebrate, it should be noted that only in a world where impoverished populations are routinely denied access to comprehensive medical care is the extension of Mwanamugimu in Amin's Uganda an unequivocal mark of success.

In the midst of rapidly worsening conditions, Kakitahi and Musoke put their energies into extending the preventive promise of Mwanamugimu to a greater number of Ugandan people. One of the principal ways that they sought to accomplish this was by training additional personnel. The public health workers or "nutrition scouts" that they trained were people who, like "Ssalango" Stephen Maseruka Mulindwa, were specifically chosen by their communities. After six months of training in the basic principles of community-oriented primary health care, they were then given a bicycle and small drug kits containing basic medicines, like antimalarials and oral hydration salts.[41] They then moved from house to house within a defined area providing education, support, and basic medical care to people within their communities. Ssalango, who began his work in the area served by the Luteete Health Center in 1978, reported that one of their central responsibilities was to identify children with faltering growth. Using arm circumference to assess nutritional status, Ssalango and other community health workers, as they were also known, provided health education to the mothers and guardians of children identified as mildly malnourished. Moreover, not unlike the influential women trained at Mwanamugimu, they were able to teach people how to prepare kitobero in their own homes using foods that were already available to them and were already an integral part of their daily meal preparation. In addition to this early intervention, community health workers were also well positioned to conduct follow-up visits in order to ensure that children identified as mildly malnourished did not continue down the "road to kwashiorkor."[42]

This further, albeit limited, expansion of medical provision into rural areas was possible during Amin's dictatorship in a way that may not have been true in urban regions of the country. Antipathy toward the educated elite meant urban centers bore the brunt of the violence perpetrated by Amin's forces. Rural communities, particularly in Buganda, remember Amin's rule as a break from the violence they faced immediately before and directly after Amin. The people living near the Luteete Health Center who shared their stories of this period with me spoke mainly of resource scarcity and having to wait in long lines for commodities like sugar and salt.[43] Expanding preventive health services into even a small selection of rural communities during a period of pronounced political violence may not have been possible, however, if it were not for the rural health services previously established at health centers like Kasangati and Luteete. What Kakitahi and Musoke were able to achieve by building on the foundations of earlier public health programming and medical services demonstrates just what long-term investments in infrastructure and expertise can bring.

Due to the economic devastation during Amin's dictatorship, the extension of the program even in the rural context was initially limited to the health centers in Luteete, Kasangati, and Kayunga. Economic embargoes compounded the impact of the oil crises and Amin's "economic war," crippling the Ugandan economy. Government price controls led many to turn to smuggling, creating a parallel economy that diminished government revenues, further weakening the state.[44] Physicians facing the decreasing value of their salaries established private practices and pharmacies.[45] Many also coped with the substantial decline in real income by shifting from cash to consumption crops in order to ensure a modicum of food security.[46] Plunging coffee prices in the late 1970s further threatened Amin's tenuous hold on power, and in 1979 Tanzanian forces marched on Kampala, forcing Amin and his men to flee into exile.[47] The people of Uganda flooded into the streets welcoming their "Tanzanian liberators" and, despite the heightened levels of violence and looting that followed, hopes ran high that the interim government would restore stability and pave the way for a period of reconstruction and prosperity. Sadly, this was not to be.

In the years immediately following Amin, Kakitahi set to work rejuvenating Mwanamugimu and extending the nutrition scout program even farther afield. Rebuilding the damaged facilities, restoring all aspects of the Nutrition Rehabilitation Program, and expanding outreach required funding. In 1983, Kakitahi secured a one-time $245,000 "Health, Hunger and Humanity" grant from Rotary International and, after reinstating all aspects of the central Mwanamugimu Unit at Mulago, including importantly the training of medical personnel, embarked on what he referred to as the "second phase." The second phase centered on first resurrecting the nutritional programs at the initial satellites in Kayunga, Luteete, and Kasangati. By 1984 they were reportedly serving an estimated 650,000 people.[48] Kakitahi then turned his attention to the establishment of additional outreach programs at health centers in Mbale, Jinja, Bugembe, Kitovu, Imbanda, Kabale, Kisomoro, and Gulu—although the latter required an intermediary due to heightened insecurity in the North. These eleven satellites of Mwanamugimu then became the foundation of a national program that Kakitahi called the Mwanamugimu Nutrition Services.[49] Kakitahi envisioned a public health program that could promote the health and well-being of a newly constituted national public. When a representative from Rotary International visited Uganda in 1985, they left so impressed with the program and its positive local reception that the organization decided to make a rare exception to their policy of only dispersing one-time funding. Kakitahi assumed that this would indeed be the final grant from Rotary International and stretched this second disbursement over a three-year period. Then, after Rotary International visited again in 1987, another exception was made and another grant issued—bringing the total funding to $980,000 or nearly one million dollars from 1983 through 1992.[50]

What is more, Kakitahi accomplished this restoration and expansion of the Mwanamugimu Nutrition Services in the midst of continued political crisis and upheaval. In 1980, forces loyal to Obote overthrew the interim government and, in order to create the façade of democratic legitimacy for his regime, Obote held elections in which outright coercion and fraud assured an electoral victory for himself and his party.[51] Obote's second term in office, known as Obote II, met with immediate resistance in several different parts of the country. The guerrilla war, or Bush War, conducted by Yoweri Museveni's National Resistance Army (NRA) in a region of central

Buganda that became known as the Luwero Triangle proved to be the most protracted of the armed movements opposing Obote's regime. The NRA depended on civilian support so Obote's army responded by terrorizing civilian populations.[52] More than eighty thousand were imprisoned in concentration-like camps and many were massacred, with the estimates of the lives lost during Obote II ranging from three hundred thousand to one million.[53] Growing international awareness of these atrocities cost Obote the international support on which his power depended. In 1983, in an apparent effort to restore his international image, Obote allowed a number of international institutions to begin providing relief to the displaced people in Luwero and other districts. Then in 1984 the government began disbanding the camps, creating a refugee crisis, as few could safely return to their homes. Vastly inadequate water and food supplies combined with overcrowding and a lack of sanitation left people in both camps and transit centers facing starvation and disease.

By the middle of 1984, the Red Cross, UNICEF, Oxfam, and the Save the Children Fund (SCF) were organizing food, water, and medical assistance for more than 150,000 people.[54] To combat the starvation and malnutrition facing many in Luwero, Oxfam and SCF set up supplementary feeding centers in a number of camps in June 1983 and "therapeutic feeding" in the government facilities at Mulago, the Luwero Health Center, and the Luteete Health Center. The nutrition scout Ssalango temporarily applied his expertise in community health work to the position of emergency relief worker, and the Luteete Health Center became a temporary feeding center where people were brought when, as Ssalango explained, "they were in a bad condition having malaria, malnutrition and other complications with regards to health."[55] The expertise and infrastructure at Luteete were, therefore, instrumental to the relief effort, providing essential care and life-saving treatment to a significant number of people. The closure of the Bombo road in March 1984 prevented food supplies from reaching an estimated 50,000 people and may have resulted in a substantial number of deaths among the 700 children served by just one of these feeding centers.[56]

In the end, the NRA forces succeeded in ousting Obote, capturing the capital city in 1986. Under the leadership of Museveni, the National Resistance Movement inaugurated a long period of political and economic stability in southern and central Uganda.[57] The intervening period of civil war had been devastating and, coming on the heels of Amin's brutal dictatorship, left the NRM with a significant task of recovery and rehabilitation. The consequences for health and well-being were great. Despite the lack of reliable data, A. B. K. Kasozi cites an estimated six-year drop in average life expectancy and a rise in the infant mortality rate from 91.9 per thousand in the early 1970s to more than 104 per thousand in 1988. Aggregate under-five mortality rates rose to as high as 172 per thousand, with the infant mortality rate among displaced peoples in the Luwero Triangle estimated at approximately 300 per thousand. Moreover, the ratio of doctors fell from one physician for every 10,000 people to one doctor for every 25,000. As Kasozi notes, "Not so long ago, Uganda had one of the best health-service delivery systems in sub-Saharan Africa. . . . This delivery system was supplemented by good preventive systems in the country. However, as a result of prolonged social conflict, both systems fell apart."[58] But the work of Kakitahi, Musoke, Ssalango, and the dedicated staff at Mwanamugimu suggests that not everything fell apart. Building on existing infrastructure, their expertise, and their capacity to garner scarce resources, these medical personnel, together with the mothers and influencers trained at Mwanamugimu, continued serving the Ugandan people, maintaining and even extending public health services and programming against all odds.

In the Shadow of Structural Adjustment

Like many of their colleagues across the continent, in the 1980s, Ugandan authorities began to enact economic reforms with significant consequences for public health and well-being. Precipitated by the economic straits following the 1973 oil crisis and world recession, these reforms involved a package of fiscal policies designed to open markets and cut government spending. In the 1960s the Ugandan economy had been particularly robust, boasting annual growth rates that exceeded demographic expansion, but by the end of Amin's dictatorship it was in ruin. General scarcity and surging rates of inflation, unemployment, and smuggling led to an expanding black market that further compounded mounting external debts as well as government and balance-of-payment deficits. When Obote returned to power after the fall of Amin, he turned to the World Bank and International Monetary Fund (IMF) in order to stabilize the situation and finance the consolidation of his power. Obote reportedly signed the agreement that became Uganda's first structural adjustment program (SAP), despite the fact that negotiations were ongoing and had apparently stalled over Ugandan concerns surrounding IMF conditionalities.[59] In exchange for agreeing to stringent economic reforms focused on long-term macroeconomic growth and market liberalization, Obote's government received substantial funding from the IMF, the World Bank, and other foreign donors. Early economic gains were ultimately reversed, however, by the Bush War that drove Obote from office in 1986 and plummeting coffee prices. With total foreign debt increasing from $733 million in 1981 to $1,031 million in 1985, Uganda was left with a staggering burden of debt and little if anything to show for it.[60] When Museveni took power the following year, inflation increased precipitously by 237 percent and, with the fall in world coffee prices, roughly four fifths of Uganda's foreign exchange earnings went to servicing the country's ballooning debt obligations.[61]

Museveni initially resisted pressure to adopt neoliberal market reforms, seeking instead to return to the state controls of the prosperous 1960s. As the economy rapidly deteriorated and alternate sources of funding proved elusive, his government abandoned its Marxist ideals and began negotiations with the IMF. In 1987 the Ugandan government announced a comprehensive IMF "Economic Recovery Package," yet when the government's subsequent budget was at variance with the agreement, further negotiations were promptly held in Washington, DC, convincing Ugandan officials to draw up an entirely new budget. Only then did the IMF, acting as lender and gatekeeper, begin to furnish substantial funds, with the World Bank and other donors following suit.[62] At the outset the government's democratic reforms appeared promising, yet before long an increasingly autocratic regime was presiding over a country that by macroeconomic standards was performing quite well. Foreign aid accounted for a growing share of the government's budget, and international financial institutions' and foreign donors' direct influence on policy increased.[63] Nor did this change in the late 1990s and early years of the new millennium, when, in response to mounting evidence that SAPs negatively impacted poor populations, the IMF and World Bank replaced structural adjustment with so-called "poverty reduction" measures.[64] Although developed by national governments ideally in consultation with civil society organizations, the supposed ownership and restoration of national sovereignty was largely empty rhetoric as firm budgetary restrictions and IMF review and approval requirements remained in place.[65]

Widespread criticism of the austere "belt-tightening" measures of structural adjustment in Africa has long centered on the mandated cuts to social services, including education and health. In Uganda the picture is a bit more complicated, suggesting that the impact of imposed neoliberal reforms cannot be reduced to a simple calculus of declining expenditure. The virtual collapse of the health system under Obote and Amin meant the budget had nowhere to go but up. According to one estimate, the Ministry of Health spent 85 percent less per person in 1982 than ten years prior, and so Uganda perhaps alone saw public health spending increase under structural adjustment.[66] What structural adjustment meant for Uganda was the loss of sovereignty over health policy and provision. Externally imposed funding restrictions impeded the reconstruction and restoration of a health system that had once been the envy of an entire continent.[67] Alongside an underfunded and increasingly demoralized national system, health projects and programs developed and funded by international organizations and NGOs proliferated. The result was a "fragmented" and "projectified landscape," creating—as in other regions of the continent—parallel and highly uneven systems of care.[68] Moreover, this process expanded significantly as a result of growing international interest in HIV research and programming.[69]

This loss of sovereignty over health policy and planning became apparent very early on. In 1987 the government created a Health Policy Review Commission composed predominantly of Makerere-trained physicians, including Namboze. The commission toured the country assessing the state of the health system and drew up plans for the restoration of basic medical provision and care. Their proposals called for the rehabilitation of infrastructure, including reconstruction of damaged facilities that, like Mulago Hospital, had been looted by Amin's troops in 1979 as they fled the capital city, as well as those that had fallen into a state of critical dilapidation after years of neglect. For many of these facilities, basic services such as water, sanitation, electricity, transportation, and drug provision also had to be restored. The commission stressed the need to establish a primary health care system and, in particular, resume and extend immunization programs. The costs of the commission's proposals reportedly "prompted a major outcry" among the donor community, and international advisors then became directly involved in the development of future health policies. The three-year health plan released in 1992 was, for example, predominantly the work of foreign experts and notably called for a hiatus on infrastructure expansion and for services to be determined by "available resources."[70] In another instance, the World Bank refused to issue a loan for the Community Health and AIDS Project until user fees were implemented, despite the fact that the democratically elected Ugandan legislature had rejected a law mandating user fees in 1990. From 1992 until they were abolished by presidential decree in 2001, patients were required to pay for services at government hospitals and health centers regardless of whether or not they could afford to pay.[71] In another instance, a European drug agency sought a monopoly over drug provision and sales in Uganda, and when the legislation establishing their monopoly stalled in the legislature in 1993, the agency reduced drug supplies to Uganda by 60 percent. Despite public outrage, the proposed bill was subsequently pushed through and, with a few amendments, later enacted into law.[72]

In the context of this growing international influence at the policy level, doctors sought to provide and patients sought to obtain quality medical care in Uganda. On top of declining infrastructure, shortage of personnel, and perpetual lack of essential drugs, medical professionals also continued to work for exceptionally low salaries. In the mid-1980s, in the context of intense inflation, medical personnel earned between $2 and $12 per month, and while this increased for medical officers to $16 in the late 1980s and then far more substantially to between $90 and $150 in 1994, doctors' salaries remained insufficient to meet their financial obligations, especially their need to provide education and housing for their families.[73] In fact the historian John Iliffe has even concluded that doctors' "loss of status and income was more extreme . . . than that of teachers."[74] The mandated decentralization of medical care that began in 1993 further exacerbated pay scale discrepancies and inconsistent salary disbursements, with some medical personnel going for extensive periods without receiving a paycheck. Medical personnel responded to these circumstances by continuing to supplement their low wages through private practices and informal charges for services, as they had done under Amin. Thus both physicians and scarce drugs continued to be found in private clinics and pharmacies rather than hospitals and health centers, even long after the restoration of peace and prosperity in southern Uganda.[75] Those seeking care turned to private services, local healers, and self-medication—even preferring to give themselves injections using their own needles and syringes.[76] In this light, Mwanamugimu represented a public health program with much to offer. It was not only an exceptionally low-cost program of rehabilitation and prevention, but, by emphasizing the preparation of a nutritious meal composed only of foods already available within the home or garden, it easily fit within the practice of self-medication that emerged in response to the ongoing decline of health provision and care in Uganda.

Ugandans also began to rely on a patchwork of programs and services developed and provided by international agencies and nongovernmental organizations. These NGOs and global health institutions filled the vacuum created by the national health system shrinking under structural adjustment. The policy of the United States President's Emergency Plan for AIDS Relief (PEPFAR) was, for instance, to provide funding for a number of distinct projects and organizations ranging from NGOs, to faith-based groups, to parastatals. Even though it meant diminished sovereignty over health, the Ugandan government had much to gain from the increased stability and state legitimacy that came with this provision of services. As Susan Reynolds Whyte and her colleagues have argued, "donor dependence (or donor mobilization) became a strategy to strengthen the state."[77] As the economy recovered, the government chose to invest its limited resources in security and further economic growth rather than health care, as the latter could more readily draw foreign aid. In the context of the international fervor for primary health care, for instance, the Ugandan government strongly supported primary health care initiatives as doing so, in Iliffe's estimation, "promised to open NGO and donor agency safes."[78] Already by the mid-1980s, NGOs delivered approximately half of all medical provision in Uganda, and in 1995 donor funding accounted for an estimated 80 percent of the Ministry of Health's budget.[79]

Medical personnel responded to the increasingly fragmented and projectified landscape of care in Uganda through protests, strike movements, and leaving altogether. In the late 1980s and early 1990s, a series of strikes and protests held to oppose the implementation of structural adjustment were met with military force. During a protest against the imposition of cost sharing at institutions of higher learning, two Makerere students were shot and killed. In 1995, medical professionals across the country participated in a ten-day strike that crippled health services and forced the government to secretly try to negotiate a settlement, with "living wage" increases ultimately only provided to a select few.[80] By the early 1990s, estimates indicate that over six hundred Ugandan doctors were working overseas. Not all left the continent, however. Namboze joined the World Health Organization as the first WHO representative to Botswana and later became the director of support for health services development at the WHO's regional office in Congo-Brazzaville.[81] Kakitahi, who had been working to extend the Mwanamugimu nutrition services to outlying districts became, in his words, "frustrated" when he could no longer secure the necessary funding. After learning that he "didn't have any money to continue," colleagues working at the WHO invited Kakitahi to develop a training program for health workers in newly independent Namibia.[82] Kakitahi's departure out of frustration with the lack of available resources under structural adjustment marks a turning point in the history of nutritional programming in Uganda. It is also a telling indication of just how devastating IMF and World Bank policies were to Ugandan health and welfare. Kakitahi returned to Uganda during the height of Amin's brutality and remained through the tumultuous insecurity of Obote II, when hundreds of thousands of Ugandans lost their lives in a brutal war. To carry on under such circumstances only to leave in the face of the obstacles of structural adjustment speaks to "the demoralization of health workers" that, as many observers note, became the principal way that Ugandans described their national health system in this period. Kakitahi remained in Uganda through Amin and the Bush War only to flee the imposition of structural adjustment.

In the Shadow of HIV

Before Rotary International issued Kakitahi with a "one-time" grant for the third time, there had been little hope of securing additional funding from that particular source. The three-year health plan issued in 1992 meant that within the context of structural adjustment, state funding for Mwanamugimu was equally unlikely. The possibility of procuring external aid from other international organizations was also doubtful, for childhood malnutrition was no longer a central focus of international concern as it had been in earlier decades. A significant new threat to health and welfare, particularly in Africa, was, for good reason, rapidly stealing the limelight and the lion's share of the resources. HIV appears to have spread from West-Central Africa into the East African region in the 1970s. Following the commercial and transportation networks forged by the smuggling or *magendo* economy that emerged under Amin, the first rural AIDS epidemic in Africa erupted in southwestern Uganda in the 1980s. The intense insecurity, troop movements, and associated sexual violence of Amin's dictatorship, the Tanzanian invasion, and the Bush War helped spread HIV to Kampala and other regions of the country. By the late 1980s prevalence rates among pregnant women in Kampala were—at 24 percent—higher than anywhere else in the world with only one exception.[83]

After two decades of violence and warfare in Uganda and the constraints of structural adjustment, the national health system was in no position to cope with an epidemic of such magnitude. Donor interest in HIV/AIDS work in Uganda was furthered, moreover, by the crucial research carried out by Ugandan medical professionals early in the epidemic, and especially by Museveni's open approach to and honest assessment of the problem and its prevention. Uganda launched—with WHO assistance—the first national control program, which was then hailed as a model, and at a conference in Kampala donors pledged over $20 million in support. The World Bank then convinced the government to create a Uganda AIDS Commission and, although competition over resources continued among the Ministry of Health, other government bodies, and different donors, international interest in funding HIV/AIDS work in Uganda grew precipitously.[84] Nongovernmental organizations led the way with more than six hundred NGOs engaged in AIDS-related activities in Uganda by 1992—a figure that soared to two thousand by 2003. The AIDS Support Organization (TASO), founded by Noerine Kaleeba in 1987, became the most well-known NGO in Uganda. After losing her husband to AIDS, Kaleeba and her colleagues worked to fight stigma and to help people "live" with AIDS by providing essential counseling and assistance to those living within the limited region where TASO operated. Like many of the NGOs working to alleviate the suffering associated with the epidemic, the majority of TASO's funding came from foreign assistance, with 97 percent supplied by international donors in 1999.[85]

Until 2004 the antiretroviral drugs that were found in 1995 to be an effective AIDS treatment were largely unavailable in Africa. Even when Indian pharmaceuticals began producing far cheaper generics in the early years of the new millennium, they remained too costly for the large majority of Ugandans with HIV. The establishment of PEPFAR and the Global Fund to Fight AIDS, Tuberculosis and Malaria substantially increased the international aid for HIV/AIDS initiatives in Uganda and other regions of the continent and shifted the emphasis of AIDS work from prevention and control to the provision of treatment. With the Gates Foundation and other donors also contributing, international funding for HIV/AIDS work around the world increased threefold to $6.1 billion per year, and this aid provided an estimated 50 percent of all HIV/AIDS-related expenditures in Africa. The extension of treatment also fueled a sharp increase in HIV/AIDS-related research on the continent, including vaccine development and trials—with Uganda serving in 1999 as the site of Africa's first vaccine trial. In Uganda the bulk of this research has focused on treatment, as the relatively recent access to antiretroviral therapies makes the Ugandan population a particularly valuable subject of HIV/AIDS-related scientific inquiry.[86]

To facilitate this HIV/AIDS research boom in Uganda, an American AIDS researcher convinced the pharmaceutical company Pfizer to contribute $5 million toward the construction of an Infectious Diseases Institute (IDI) on Mulago Hill. As the first new construction within the Makerere-Mulago complex in thirty-seven years, the IDI laboratory, with its "bright exterior," contrasts sharply with "the sooty, weathered façade" of the surrounding hospital buildings. As Johanna Crane has insightfully observed, the IDI thereby serves as "an architectural testament to the special access to donor money that HIV holds over other afflictions."[87] Final construction of the institute and its grand opening in 2004 coincided with my longest period of research in Uganda for this project, research involving repeated visits to the Mwanamugimu Nutrition Rehabilitation Unit located a bit further up the slopes of Mulago Hill. At the time, the bustling activity around the institute made the quiet and aged look and feel of Mwanamugimu that much more evident. Even though women and their children attended outpatient clinics and gathered at mealtimes to prepare and consume kitobero, Mwanamugimu appeared somewhat empty and run-down. A great deal of my first meeting with the unit's director, Elizabeth Kiboneka, was spent outlining the laundry list of items that the unit required in order to function at capacity and as intended. In particular the unit no longer had access to a vehicle and—clearly hoping that I might be of some use in securing the required donor support—she explained how essential transportation was to follow-up and outreach. Thus while Mwanamugimu desperately sought to acquire a single vehicle, state-of-the-art laboratory equipment was transforming a small part of Mulago Hill into an international hub of HIV/AIDS funding and research. The IDI was so advanced that it became the only laboratory in East Africa with the certification verifying international standards of quality control, which, as Crane notes, meant that it "was preferred by major funding agencies such as the [National Institute of Health] NIH and its British equivalent, the Medical Research Council."[88] In other words, the very organization that once funded cutting-edge nutritional research on Mulago Hill now supports HIV/AIDS work, and Uganda has again become a center of global health research and programming. The Infectious Diseases Institute thus stands as a stark reminder of what Uganda's National Institute of Human Nutrition might have been.

In 2011, Mwanamugimu received funding for a fresh coat of paint, the acquisition of new beds and equipment, additional staff, and food. Painted in bright nursery colors and adorned with cartoon characters, the unit now resembles a typical nursery or pediatric hospital ward. According to Kiboneka, the donated equipment has been particularly useful in reducing mortality at the unit. During my visit in 2012, the transformation was clear. Although the upper buildings used for demonstrations had not been updated and appeared to be no longer in use, far more children and their parents were attending the unit and especially the newly repainted wards specializing in medical treatment and care.[89] In addition to the involvement of the Christian nonprofit RESTORE Uganda, the Shs. 380 million for Mwanamugimu's "facelift" were raised through the charitable donations of "friends and fellow churchgoers" of a United Kingdom woman connected to Mildmay Uganda—a faith-based HIV treatment center funded by PEPFAR.[90] Here international HIV/AIDS funding provides opportunities for further charitable work, or what might be thought of as an indirect form of "trickle down" global health funding and support. One online comment on Mwanamugimu's facelift from "Ugandanmedic," speaks to the tragedy that this represents when she asks her fellow Ugandan citizens, "Are we not embarrassed that a foreigner can do for us what the government is supposed to do?"[91]

The cruel irony of a nutrition rehabilitation program struggling to continue in the shadow of HIV/AIDS funding and research is the false separation of poor nutritional health and HIV infection. Not only are the populations facing food insecurity and undernutrition also those with the highest burden of HIV, but recent research has shown a number of "negative feedback loops" in which HIV significantly increases malnutrition and poor nutritional health considerably aggravates HIV—potentially accounting for a significant component of high HIV morbidity and mortality even in the wake of greater access to treatment and care. In addition to increasing recognition of the "nutrition-infection complex"—to use Whitehead's earlier terminology—and its relevance to HIV and nutritional health, there is also a growing awareness that the hunger associated with HIV treatment creates significant problems for households and individuals with insufficient access to food and other resources.[92] As a result, international organizations have recently begun to argue that "targeted nutritional interventions should be systematically linked to antiretroviral interventions," and "the World Bank has called for a scaling up of action on nutrition and AIDS through 'action research' and 'learning by doing.'"[93] One analysis of "HIV programs in Africa concluded that current HIV/AIDS policies 'have tended toward highly medicalized approaches' and called for 'a comprehensive approach to link health strategies with community-oriented food-based strategies.'"[94] In light of the nutritional work carried out in Uganda for the better part of a century, it is hard not to see this emerging awareness as both an inherently positive development and somewhat sad reinvention of a wheel originally devised more than forty years ago. Moreover, as early as 1993, experts estimated that up to half of children suffering from severe malnutrition in some countries were HIV positive.[95] Malnourished children with HIV can be treated even without access to antiretrovirals, but their recovery takes longer and is more likely to have complications and result in death.[96] Over the years a growing number of the severely malnourished children brought to the Mwanamugimu Unit have been HIV positive and, in 2002, prior to the rollout of antiretrovirals in Uganda, the staff at Mwanamugimu expressed deep disquiet at their inability to fully treat and rehabilitate these young children. Many of those who survived then joined the ranks of those known as AIDs orphans, as they lost their parents to the disease.

Mwanamugimu has thus come to serve an additional purpose—providing knowledge of how to promote nutritional health to an increasing number of grandparents who are now caring for children orphaned by AIDS. Observers have long noted that the burden of care for both people dying of HIV infection and the children they leave behind has fallen on families and especially women—with an increasing number of grandparents responsible for raising their young grandchildren. A number of the elderly women and men interviewed for this study were grandparents who had outlived their own children, and were now caring for the grandchildren left behind. Nabanja Kaloli's son, a school teacher who had been present during my first interview in 2004, died only a week prior to my visit in 2012, leaving Nabanja, as a widow, with sole responsibility for a number of her grandchildren. Although not all of these children were AIDS orphans or orphans at all, it was nonetheless clear that, for those that were, the Mwanamugimu program had been instrumental to their survival and care. Maria Zerenah Namusoke told me about an infant that had been put in her care when the child was only one month old and, according to her testimony, she "prepared for this child kitobero" and the child "is in good condition because of kitobero."[97] Peragiya Nanziri explained that after "her girl died and left her with a nine-month old baby," she "prepared kitobero for the baby and now he is healthy and still alive."[98] One grandmother indicated that she "prepares kitobero for her grandchildren . . . because she is the one taking care of those babies."[99] When Dorobina Kyambadde began in 2005 to care for "the child of her daughter, who was suffering from malnutrition," she took the child to the Mwanamugimu Unit on Mulago Hill and the kitobero she then fed her "helped the child to grow well."[100] Thus even with very limited financial resources at its disposal, Mwanamugimu has continued to serve the severely malnourished children within reach of the capital city. Moreover, those who have made the program part of their living memory and social practice, as in the region surrounding the Luteete Health Center, have found their knowledge of how to prepare kitobero critical as they cope with the growing responsibilities of grandparents within the fragmented health system, combating an epidemic of HIV.

It is difficult to conclude this chapter with dispassionate analysis. Instead, I am left with a mixture of anger and awe. What John Kakitahi, Latimer Musoke, Ssalango, and Gladys Stokes and her team achieved even in the midst of the terrible violence and upheaval perpetrated by Amin and Obote, is, very simply, awe-inspiring. That the neoliberal policies of structural adjustment converged with global health funding faddism and faith in the panacea of nongovernmental organizations to make Mwanamugimu dependent on a "trickle-down" form of philanthropy for essential funding is hard to swallow. But a number of important insights follow from a close look at the postcolonial history of this innovative public health program. First and perhaps foremost, Mwanamugimu could be sustained through the combined onslaughts of Obote, Amin, and global health in the neoliberal age because it became a thoroughly local program, dependent on local advocacy, expertise, and infrastructure more than on the shifting donor interest and support that has become essential to so much global health work in Africa. In this light, what is clear is that Mwanamugimu endured because it built on the robust foundations of biomedical training, expertise, and infrastructure—on the robust foundations of a national health system. As a result, Mwanamugimu had the capacity to become a fully local initiative and this, combined with the clear value of the program itself, largely accounts for its remarkable longevity. What Kakitahi, his colleagues, and the influencers who kept Mwanamugimu alive illustrate is the return on long-term investments in national systems of medical provision, which can then serve as the foundation for flexible and resilient public health programming, programming that is capable of promoting health and wellbeing in ways that increase rather than undermine sovereignty and the right to health that all people deserve.

EPILOGUE

Remedicalizing Malnutrition and the Plumpy'Nut Revolution

Severe acute malnutrition has recently returned to the international limelight. Several factors have played a part in putting the condition back on the global agenda, including the attention to hunger and child mortality within the Millennium Development Goals adopted by the United Nations in 2000.[1] Yet prior to 2005, severe acute malnutrition was not specifically the focus and, despite its importance in under-five mortality, was often entirely overlooked.[2] All that changed, however, when the NGO Médecins Sans Frontières (MSF) launched a massive and highly effective treatment campaign in Niger using a new therapeutic regimen that revolutionized the treatment and prevention of severe malnutrition in young children. The promise of this innovation has galvanized a new generation of experts, NGOs, and manufacturers to keep the specter of severe childhood malnutrition at the forefront of global health. There is no question that this therapeutic innovation plays an invaluable role in war-torn and crisis situations. What is not clear is whether reliance on this magic-bullet solution with striking similarities to the efforts of the 1950s and 1960s is, in fact, the key to prevention. The remedicalization of malnutrition that it may represent appears to favor a foreign aid–based humanitarian approach and a commercial, drug-based model that, if the past is any guide, could leave future generations more rather than less vulnerable—especially when compared with a more enduring, cost-effective, and comprehensive public health model like Mwanamugimu.

The therapeutic innovations that returned severe malnutrition to the spotlight have a genealogy reaching back to the 1970s and 1980s when Michel Lescanne first began working to develop "a nutritional biscuit for populations in developing countries." In 1976, Lescanne helped to create a nutritional bar known as Novofood, and in the early 1980s Novofood was distributed in the Karamoja region of northeast Uganda by an NGO working to combat a famine that reportedly took the lives of more than 20 percent of the population in just twelve months.[3] In 1984, in what became known as Operation Sahel 84: Trucks of Hope, the French Red Cross distributed 6 million Novofood bars to famine-stricken populations stretching from Mauritania to Niger.[4] Lescanne then launched a company called Nutriset with a stated mission to conduct "research in the field of humanitarian nutrition, developing innovative solutions and acting as an interface between the worlds of humanitarian aid, nutritionists and food industry technologies."[5] This work finally bore fruit in the early 1990s when Lescanne teamed up with two nutritionists, André Briend and Michael Golden. Nutriset's first products focused on simplifying the treatment of severe malnutrition by NGOs, yielding a "ready-to-dilute therapeutic milk" known as F-100 followed by F-75, with the latter specially formulated for the early stages of treatment. These mixtures of dried skim milk, oil, and sugar with added vitamins and minerals very closely resembled and were essentially the commercial equivalent of the original kwashiorkor treatments developed by Dean in the 1950s.[6] Following successful trials in Rwanda in 1993, these therapeutic milk powders and other products designed to improve the treatment of severe acute malnutrition in young children were embraced by organizations engaged in nutritional therapy around the globe and endorsed by the WHO in 1999.[7]

Treatment using therapeutic milks requires reconstituting the powder in water to form a liquid diet that, as a result and as we have seen, has a number of disadvantages. In particular, the liquid formula is especially prone to bacterial contamination and must therefore be prepared immediately prior to feeding and by trained personnel. As a result, the treatment of severe malnutrition using these therapeutic milks requires inpatient facilities, access to clean water, and expert care. The length of treatment means that the number of children who can receive therapy at any one time is necessarily limited by the capacity of hospitals and health centers.[8] As Nutriset's official website explains, Lescanne and Briend thus set out to develop a substitute for F-100 and F-75 in order to "simplify the work of NGOs" by developing "an alternative . . . more suitable for the conditions of use on the ground."[9] After failed attempts to devise a viable pancake, doughnut, or biscuit, Briend, reportedly inspired by the hazelnut-based paste Nutella, seized on the idea of using a spread in which the F-100 formula remained virtually the same, but was mixed with peanut butter to form what became known as a "ready-to-use therapeutic food" or RUTF. Due to a far lower moisture content and less contact with the air, RUTF has a much longer shelf life and is less prone to contamination by insects and bacteria. Initial field trials carried out in

Chad in the late 1990s found that children preferred RUTF spreads to the skim-milk liquid diet.[10]

The real breakthrough came when the Nutriset RUTF formula, patented as "Plumpy'Nut," became the basis of the first large-scale outpatient campaign to treat children suffering from severe malnutrition. The outpatient treatment of malnutrition using RUTF was first pioneered by the NGO Concern Worldwide at emergency therapeutic feeding centers in Ethiopia during the famine of 2000 and 2001.[11] Then following a visit from Briend in 2002, MSF began moving the treatment of children suffering from severe malnutrition in Niger to outpatient facilities. By 2005, the majority of the malnourished children brought to the MSF therapeutic feeding centers in Maradi, Niger, received rations of Plumpy'Nut and recovered in their own homes in the care of their mothers or guardians.[12] Treating children outside of hospital settings was not only efficient in terms of facilities and personnel, but also decreased their exposure to other infectious diseases and notably reduced the opportunity costs for caregivers who otherwise remained with hospitalized children for extensive periods, neglecting obligations at home. Outpatient treatment therefore appears to have added benefits for parents and guardians, possibly impacting acceptance and demand for RUTF and nutritional therapy.[13] When the food crisis of 2005 dramatically increased the number of children seeking treatment in Niger, MSF was able to illustrate their ability to scale up their provision of treatment, with more than forty thousand severely malnourished children treated on an outpatient basis using RUTF. Although earlier RUTF trials had already indicated the efficacy of Plumpy'Nut, the unprecedented scale and apparent success of the campaign brought significant attention to severe childhood malnutrition and the novel therapeutic foods that suddenly made widespread treatment and prevention appear feasible.[14]

The "Plumpy'Nut Revolution" and the "Peanut Butter Debate" it subsequently sparked were never solely about treatment. Rather, it was the preventive promise of Plumpy'Nut and similar products that became embroiled in ongoing controversy and dispute. Emboldened by the success of outpatient treatment in 2005, MSF then embarked on an effort to treat children who were moderately malnourished and expanded their program, distributing Plumpy'Nut to more than sixty thousand children with over 95 percent reportedly achieving a full recovery. Then in 2007, Nutriset developed "Plumpy'Doz," a similar product with a higher vitamin and mineral content designed to be eaten in relatively small amounts of approximately three tablespoons per day, as a supplement rather than a replacement for daily meals. Arguing that it made little sense to wait until children became malnourished, MSF launched a program to distribute four containers of Plumpy'Doz per month to all mothers and guardians of children between six months and three years of age within a single district of Niger. Another sixty thousand children were included in this "blanket distribution," which, according to MSF, led to a lower prevalence and incidence of severe malnutrition as compared to prior years.[15]

It was at this stage that the media joined MSF and others in proclaiming the virtues of Plumpy'Nut as the "hunger-wonder product" and "the most important weapon in the war on global hunger."[16] Following Anderson Cooper's 2007 report on *60 Minutes*, which described Plumpy'Nut as a "miraculous cure" that "may just be the most important advance ever to cure and prevent malnutrition," there was an explosion of interest in RUTF, with numerous nonprofit companies and organizations, including many faith-based groups, joining in the effort to produce and procure Plumpy'Nut for malnourished children around the globe.[17] Edesia, the first American company to begin manufacturing RUTFs as a licensed subsidiary of Nutriset, was started when, as the company's website states, "in 2007, Navyn Salem, a stay-at-home mother of four young girls, set out with a clear, yet ambitious goal to end the crisis of malnutrition for over 200 million children around the world."[18] The answer for Salem and many who have joined her was in the production and widespread distribution of RUTF, or in Salem's words, "the magic stuff" that represents, as she told those attending a three-course benefit luncheon in Manhattan, "the very simple solution" to the problem of global malnutrition.[19] The Christian missionary Mark Moore, who founded MANA as a charitable organization and nonprofit company based in Georgia and South Carolina in order to manufacture and distribute its own version of RUTF or MANA (Mother Administered Nutritive Aid), was also convinced that RUTF held the answer to the global problem of malnutrition.[20] These and other endeavors have led to vociferous disputes over Nutriset's patents and the considerable revenues that the company has made in the business of curing and preventing childhood malnutrition—amounting to $66 million in sales in 2008 alone.[21] Particularly after the WHO, UNICEF, and the World Food Program began officially endorsing RUTF as an integral part of "community-based management of severe acute malnutrition" in 2007, sales of Plumpy'Nut rose rapidly, with demand at times outpacing supplies.

UNICEF, as the largest procurer, purchased over 4,000 metric tons in 2007—a figure that soared to over 20,000 metric tons in 2010 and an estimated 54,000 for 2011 through 2012. In 2012, UNICEF purchased RUTF from over twelve different global manufacturers and seven "local" production companies operating as licensed subsidiaries of Nutriset in a number of different countries across the African continent, including Dr. Mark Manary's Project Peanut Butter in Malawi.[22]

The ethical concerns of profiting from child malnutrition were further compounded by several additional considerations, among them the absence of evidence-based data confirming RUTFs as the most effective means of preventing malnutrition. Although numerous studies indicate that ready-to-use therapeutic foods are better than nothing and more efficacious than the blended soy and maize flours distributed as a means of prevention in places like Malawi and Niger, some note that in the absence of randomized control trials, the evidence is far from definitive.[23] Beyond questions of efficacy, the issues of cost and the sustainability of adopting emergency therapeutic measures for ongoing, large-scale programs of prevention became paramount. Although the cost of outpatient treatment using RUTF represented a savings over extended periods of inpatient hospital care, the same could not be said for the blanket distribution of RUTF within regions of high prevalence involving substantial numbers of potentially malnourished children. In the end, concerns over the expense of RUTF have become critical, as the costs of treatment alone are greater than the annual per-capita health expenditures in the majority of Sub-Saharan African countries. According to one analysis, treating a malnourished child with imported RUTF cost an estimated $55—$22 if locally produced—and for children suffering from HIV infection the expense doubled.[24] Moreover, large-scale programs of prevention entail additional costs, including the medical personnel needed to facilitate distribution and monitoring.[25] In theory, local production of RUTF would not only cut costs, but also contribute to regional development. Yet in practice, most if not all of the ingredients have to be imported—especially the skim-milk powder and vitamin-mineral mixture—and thus prices remained prohibitively high for long-term general distribution to be a feasible prospect for many if not most of the countries with the highest burden of malnutrition in Africa and other parts of the world.[26] What is more, the cost-effectiveness of RUTF diminishes when the comparison shifts to programs like Mwanamugimu, which rely on locally grown foodstuffs and operate on a largely outpatient basis.

For MSF and other advocates of RUTF as both prevention *and* cure, questions of cost and sustainability are viewed as synonymous with the arguments raised against antiretroviral provision in Africa. In this light, malnourished and potentially malnourished children have a right to RUTF as both treatment and prevention, just as those infected with HIV have a right to life-saving therapies. Activism around RUTF as a preventive measure has therefore been pursued by MSF along the lines pioneered in the struggle for HIV treatment in Africa and other impoverished regions.[27] Nor is this line of thinking altogether surprising given that RUTF was initially developed as a therapeutic measure and has often been spoken of and compared with other medical cures and pharmaceutical treatments. Thus when Anderson Cooper described RUTF as a "miraculous cure" and asked the head nutritionist at MSF if it was "the equivalent of penicillin," the response was yes, that RUTF is "like an essential medicine" that "can cure . . . just like an antibiotic."[28] The director of a hospital in Haiti reportedly spoke of Plumpy'Nut as an "immunization," and this medicalization of malnutrition and reliance on a targeted, magic-bullet approach has been acknowledged elsewhere.[29] As one nutritionist told a *New York Times* reporter, "People love a silver bullet."[30]

What is not understood is that RUTF represents a remedicalization of malnutrition, one that could create relations of dependency and thereby undermine prevention, as similar endeavors discovered in Uganda nearly half a century before. The failure of the high-protein food program that prompted physicians like Dean to continue with the emergency distribution of dried skimmed milk, blurring the distinction between cure and prevention, was, as we have seen, not without certain hazards for child health and well-being. In this instance, there is very little danger that mothers of malnourished children will see RUTF as an endorsement of bottle feeding, but as Jean-Hervé Jézéquel has insightfully noted, RUTF has already been appropriated by recipient communities in unintended ways, and we know far too little about the meanings that RUTF has and will have in the social worlds of different regions. What previous experiences show is that medicalizing malnutrition led many in Uganda to view malnutrition as a disease requiring curative therapies and treatments. In placing doctors again at the forefront of treatment *and* prevention, RUTF, conceived of as a therapy rather than as a food, is the opposite of kitobero. For although both rely on mothers to administer the nutritious food to their young children, only one is prepared *by* mothers using locally available foods, and only one serves to impart the knowledge of what young children need to avoid malnutrition. Only one can continue in the shadow of war, independent of donors, and across generations. Mwanamugimu was specifically designed to demedicalize malnutrition, to decenter doctors from the treatment and prevention of severe acute malnutrition in order to empower mothers and caregivers with the knowledge needed to independently promote nutritional health in their communities, while RUTF may reestablish a relationship of dependence. Here the danger of neglecting the history of past public health initiatives becomes clear. Because Mwanamugimu remains largely forgotten outside of Uganda, despite the renewed international interest in severe malnutrition it now continues in the shadow of RUTF. Those inspired by the preventive promise of RUTF also proceed unaware of the resonance with past endeavors and the lessons that were already learned.

In the midst of interviewing a member of the Mwanamugimu staff in 2012, I noticed a stack of boxes in the corner of the office containing packets of Plumpy'Nut. When I asked about the integration of Plumpy'Nut into the Mwanamugimu program, the answer was guarded but gave the distinct impression that the staff at Mwanamugimu had little say in the matter, as the new UNICEF protocols prescribed RUTF in the treatment and prevention of childhood malnutrition. This loss of sovereignty over the ongoing development of the program seems a far cry from the time when Gladys Stokes and her team were reportedly given full creative license. When I then asked what influence Plumpy'Nut was having on mothers and their appreciation of kitobero, I was told that some mothers no longer saw kitobero as sufficient to promote nutritional health in their children and asked how they would prevent malnutrition when the Plumpy'Nut ran out. This answer recalled a document written by Paget Stanfield more than four decades ago, a document describing how, prior to the establishment of Mwanamugimu, mothers brought their malnourished children back the hospital for repeated therapy despite the distribution of skim milk powder, declaring, "'Doctor, what could I do; the milk ran out?'"[31] As this return to magic-bullet solutions threatens to further erode the program, and especially the rural outreach achieved in the 1970s and 1980s, the resulting remedicalization of malnutrition may present the most intractable challenge to the legacy of the Mwanamugimu program in Uganda. Unlike Mwanamugimu, it neither builds on nor expands biomedical infrastructure and expertise in the region; instead, it may further siphon limited public health funds from what was once an integrated and far more comprehensive system of medical care and provision.

But it does not have to be this way. RUTF has a place in the global health toolkit. In the context of large-scale warfare, refugee crises, major famines, and other disasters, RUTF represents an unparalleled opportunity to treat and prevent the severe malnutrition that inevitably occurs and to thereby save an unknown number of lives. It could also be envisioned as a first step or an integral part of establishing long-term programs of prevention—public health programs that, like Mwanamugimu, focus on education and evolving efforts to promote nutritional health through local food mixtures and through a public thereby empowered and constituted. Let us not forget that Mwanamugimu also grew out of the failure of an effort to use the cure to teach prevention. It may be possible to deploy RUTF in ways that also build foundations for systems of public health that increase capacity and enhance sovereignty—public health programs that might be able to both promote health and endure.

NOTES

Preface

1. Megan Vaughan, "Reported Speech and Other Kinds of Testimony," in *African Worlds, African Voices: Critical Practices in Oral History*, ed. Luise White, Stephan F. Miescher, and David William Cohen (Bloomington: Indiana University Press, 2001), 66–67. My methodology was also influenced by conversations with Tamara Giles-Vernick and her experience collecting oral evidence for her book: *Cutting the Vines of the Past: Environmental Histories of the Central African Rain Forest* (Charlottesville: University Press of Virginia, 2002).

Introduction

1. André Briend et al., "Putting the Management of Severe Malnutrition Back on the International Health Agenda," *Food and Nutrition Bulletin* 27, no. 3 (2006): S3; Robert E. Black et al., "Maternal and Child Undernutrition and Overweight in Low-Income and Middle-Income Countries," *Lancet* 382 (August 3, 2013): 21; and Black et al., "Maternal and Child Undernutrition: Global and Regional Exposures and Health Consequences," *Lancet* 371 (January 19, 2008): 246.

2. Maggie Black, *Children First: The Story of UNICEF, Past and Present* (Oxford: Oxford University Press, 1996); and Black, *The Children and the Nations: The Story of Unicef* (New York: Unicef, 1986).

3. The condition is also diagnosed when the weight-for-height ratio is three standard deviations below what is considered average for the respective age. World Health Organization, *Essential Nutrition Actions: Improving Maternal, Newborn, Infant and Young Child Health and Nutrition* (Geneva: World Health Organization, 2013): 31; and Briend et al., "Management of Severe Malnutrition," S3.

4. Natasha Lelijveld et al., "Chronic Disease Outcomes after Severe Acute Malnutrition in Malawian Children (ChroSAM): A Cohort Study," *Lancet Global Health* 4, no. 9 (September 2016): e654–e662, doi:10.1016/S2214-109X(16)30133-4; and André Briend and James A. Berkley, "Long Term Health Status of Children Recovering from Severe Acute Malnutrition," *Lancet Global Health* 4, no. 9 (September 2016): e590–e591, doi:10.1016/S2214-109X(16)30152-8.

5. Michael Krawinkel, "Kwashiorkor Is Still Not Fully Understood," *Bulletin of the World Health Organization* 81, no. 12 (2003): 910; and Snezana Nena Osorio, "Reconsidering Kwashiorkor," *Topical Clinical Nutrition* 26, no. 1 (2011): 10–13. Research on aflatoxins and antioxidants remain inconclusive. As in the late 1960s, kwashiorkor continues to be seen as a form of protein-energy malnutrition aggravated by infection, and the role of infection is once again seen as especially important in light of recent evidence indicating that antibiotics increased survival rates.

See, for instance, Olaf Müller and Michael Krawinkel, "Malnutrition and Health in Developing Countries," *Canadian Medical Association Journal* 173, no. 3 (2005): 279–86; and Indi Trehan et al., "Antibiotics as Part of the Management of Severe Acute Malnutrition," *New England Journal of Medicine* 368, no. 5 (January 31, 2013): 425–35, doi:10.1056/NEJMoa1202851.

6. Zulfiqar A. Bhutta et al., "Evidence-Based Interventions for Improvement of Maternal and Child Nutrition: What Can Be Done and at What Cost?" *Lancet* 382 (August 3, 2013): 48–49; interview by the author with Dr. Roger Whitehead, Kampala, Uganda, December 9, 2003; and John Iliffe, *The African Aids Epidemic: A History* (Athens: Ohio University Press, 2006).

7. Black et al., "Maternal and Child Undernutrition and Overweight," 15; and WHO, *Essential Nutrition Actions*, 3.

8. Black et al., "Maternal and Child Undernutrition and Overweight," 21–26.

9. James L. A. Webb Jr., *Humanity's Burden: A Global History of Malaria* (New York: Cambridge University Press, 2009). This is also true of under-five mortality: Black et al., "Maternal and Child Undernutrition," 244.

10. Black et al., "Maternal and Child Undernutrition and Overweight," 23–26.

11. FANTA-2., *The Analysis of the Nutrition Situation in Uganda*. Food and Nutrition Technical Assistance II Project (FANTA-2) (Washington, DC: Academy for Educational Development, 2010), http://www.health.go.ug/hmis/public/nutrition/Uganda_Nutrition_Situation_Analysis.pdf.

12. I. H. E. Rutishauser, "Statistics of Malnutrition in Early Childhood (with Reference to Uganda)," Monograph No. 13 *Journal of Tropical Pediatrics and Environmental Child Health* 17, no. 1 (March 1971): 15.

13. There are a number of problems with assuming a one-to-one correlation between African and biomedical diagnostic categories and especially in this case, as it is clear that the various conditions equated with kwashiorkor were often locally defined according to the context of the child's illness rather than a specific set of symptoms. Julie Livingston, *Debility and the Moral Imagination in Botswana* (Bloomington: Indiana University Press, 2005); Steven Feierman, "Explanation and Uncertainty in the Medical World of Ghaambo," *Bulletin of the History of Medicine* 74 (2000): 317–44; Feierman, *Peasant Intellectuals: Anthropology and History in Tanzania* (Madison: University of Wisconsin Press, 1990); and F. J. Bennett, "Custom and Child Health in Buganda: V. Concepts of Disease," *Tropical and Geographical Medicine* 15 (1963): 148–57.

14. There are two interrelated sets of assumptions underlying perceptions of a largely unchanging global prevalence of severe acute malnutrition. The first is that syndromes like severe acute malnutrition cause poverty and underdevelopment and are thereby part of what accounts for the fact that many populations appear to be stuck in "cycles of poverty." At times this understanding has converged with and reinforced an environmentally determined view of "tropical" regions, including much of Sub-Saharan Africa, as inherently diseased and thereby inimical to progress and "civilization." It is also a perspective that rests on long-held views of Africa

as a place without a past, as a part of the world untouched, until recently, by historical change.

15. Economic factors were initially a focus of this research, but for two reasons were not ultimately examined as a part of this study. The limitations of the evidence made this line of inquiry a challenge that, while worth pursuing, would require consulting a very different set of sources. In part, this is due to the fact that, in the period under consideration, biomedical personnel firmly believed severe acute malnutrition was not connected to poverty or economic constraints and the data that they produced was therefore largely silent on this matter. The second reason for shifting my focus away from an examination of economic variables was to avoid obscuring the important story of Mwanamugimu and the Luteete Health Center. The role of impoverishment in malnutrition in Africa has also been addressed by prior scholars. See for example: Steven Feierman, "Struggles for Control: The Social Roots of Health and Healing in Modern Africa," *African Studies Review* 28, no. 2/3 (1985): 73–147; Diana Wylie, "The Changing Face of Hunger in Southern African History, 1880-1980," *Past and Present* 22 (1989): 159–99; Wylie, *Starving on a Full Stomach: Hunger and the Triumph of Cultural Racism in Modern South Africa* (Charlottesville: University of Virginia Press, 2001); Cynthia Brantley, *Feeding Families: African Realities and British Ideas of Nutrition and Development in Early Colonial Africa* (Portsmouth, NH: Heinemann, 2002); Henrietta L. Moore and Megan Vaughan, *Cutting Down Trees: Gender, Nutrition, and Agricultural Change in the Northern Province of Zambia, 1890-1990* (Portsmouth, NH: Heinemann, 1994); Elias C. Mandala, *The End of Chidyerano: A History of Food and Everyday Life in Malawi, 1860-2004* (Portsmouth, NH: Heinemann, 2005); and especially for Uganda: Jan Kuhanen, *Poverty, Health and Reproduction in Early Colonial* Uganda (Joensuu, Fin.: University of Joensuu, 2005). For recent acknowledgement of the centrality of political, economic, and social factors see: Richard Horton, "Maternal and Child Undernutrition: An Urgent Opportunity," *Lancet* 371 (January 19, 2008): 179; and Richard Horton and Selina Lo, "Nutrition: A Quintessential Sustainable Development Goal," *Lancet* 382 (August 3, 2013): 2.

16. Wylie, *Starving on a Full Stomach*; Randall Packard, "Visions of Postwar Health and Development and Their Impact on Public Health Interventions in the Developing World," in *International Development and the Social Sciences: Essays on the Politics of Knowledge*, ed. Frederick Cooper and Randall Packard (Berkeley: University of California Press, 1997), 93–115; and Joseph Morgan Hodge, *Triumph of the Expert: Agrarian Doctrines of Development and the Legacies of British Colonialism* (Athens: Ohio University Press, 2007).

17. Chimamanda Ngozi Adichie, *Half of a Yellow Sun* (New York: Alfred A. Knopf, 2006), 375.

18. See for example: Duana Fullwiley, *The Enculturated Gene: Sickle Cell Health Politics and Biological Difference in West Africa* (Princeton: Princeton University Press, 2011); and Didier Fassin, *When Bodies Remember: Experiences and Politics of AIDS in South Africa* (Berkeley: University of California Press, 2007). Recent scholarship has also pointed to how healers and medical professionals in Africa who are influenced

by "Western" biomedicine nonetheless interpret such influences in culturally and historically specific ways; see for example: Stacey Ann Langwick, *Bodies, Politics, and African Healing the Matter of Maladies in Tanzania* (Bloomington: Indiana University Press, 2011).

19. See for example: Melissa Graboyes, *The Experiment Must Continue: Medical Research and Ethics in East Africa, 1940-2014* (Athens: Ohio University Press, 2015); Lydia Boyd, *Preaching Prevention: Born-Again Christianity and the Moral Politics of AIDS in Uganda* (Athens: Ohio University Press, 2015); and Amy Thomas Moran, "A Salvage Ethnography of the Guinea Worm: Witchcraft, Oracles and Magic in a Disease Eradication Program," in *When People Come First: Critical Studies in Global Health*, eds. João Guilherme Biehl and Adriana Petryna (Princeton: Princeton University Press, 2013), 207–39. Also see: Shula Marks, "What Is Colonial about Colonial Medicine? And What has Happened to Imperialism and Health?" *Society for the Social History of Medicine* 10, no. 2 (1997): 205–19; and Randall Packard, "Malaria Dreams: Postwar Visions of Health and Development in the Third World," *Medical Anthropology* 17 (1997): 279–96.

20. See for example: James L. A. Webb Jr., "The First Large-Scale Use of Synthetic Insecticide for Malaria Control in Tropical Africa: Lessons from Liberia, 1945-62," in *Global Health in Africa: Historical Perspectives on Disease Control*, ed. Tamara Giles-Vernick and James L. A. Webb Jr., Perspectives on Global Health (Athens: Ohio University Press, 2013), 42–69; and James L. A. Webb Jr., *The Long Struggle against Malaria in Tropical Africa* (New York: Cambridge University Press, 2014).

21. Graboyes, *Experiment Must Continue*, 26. The notion of enduring engagements operating in this analysis is also influenced by scholarship on the importance of therapy managing groups and the emergence of "biomedical pluralism" in Africa. See Steven Feierman, "Change in African Therapeutic Systems," in "The Social History of Disease and Medicine in Africa," special issue, *Social Science and Medicine* 13B (1979): 277–84; John M. Janzen and Steven Feierman, "Introduction" in "The Social History of Disease and Medicine in Africa," special issue, *Social Science and Medicine* 13B (1979): 239–43; and John M. Janzen, *The Quest for Therapy in Lower Zaire* (Berkeley: University of California Press, 1978).

22. For the more on historical epidemiology see: Tamara Giles-Vernick and James L. A. Webb Jr., eds., *Global Health in Africa: Historical Perspectives on Disease Control* (Athens: Ohio University Press, 2013); James L. A. Webb Jr., "Historical Epidemiology and Infectious Disease Processes in Africa," *Journal of African History* 54, no. 1 (2013): 3–10; and James L. A. Webb Jr., "The Art of Medicine: The Historical Epidemiology of Global Disease Challenges," *Lancet* 385 (January 24, 2015): 322–23. For this notion of "afterlife" see Nancy Rose Hunt, *A Nervous State: Violence, Remedies, and Reverie in Colonial Congo* (Durham, NC: Duke University Press, 2016).

23. Johanna Tayloe Crane, *Scrambling for Africa: AIDS, Expertise, and the Rise of American Global Health Science* (Ithaca: Cornell University Press, 2013), 4. For more on drug trials see also Adriana Petryna, *When Experiments Travel: Clinical Trials and the Global Search for Human Subjects* (Princeton: Princeton University Press, 2009);

and Claire Wendland, "Research, Therapy, and Bioethical Hegemony: The Controversy over Perinatal HIV Research in Africa," *African Studies Review* 51, no. 3 (2008): 1–23.

24. The result might be a form of "vernacular knowledge," whereby, as Helen Tilley has argued, "insights derived from African experiences were folded into the fabric of scientific discoveries." Here the emphasis is less focused on how this process influenced science and scientific thinking, or colonial policies and critiques of colonialism, and more on how it shaped biomedical work, public health, and the resulting outcomes. Helen Tilley, *Africa as a Living Laboratory: Empire, Development, and the Problem of Scientific Knowledge, 1870-1950* (Chicago: University of Chicago Press, 2011), 319.

25. Giles-Vernick and Webb, *Global Health in Africa*, 3. See also Crane, "Doing Global Health," in *Scrambling for Africa*, 145–71.

26. For more on public health and especially in the African historical context see: Ruth J. Prince, "Situating Health and the Public in Africa: Historical and Anthropological Perspectives," in *Making and Unmaking Public Health in Africa: Ethnographic and Historical Perspectives*, eds. Ruth J. Prince and Rebecca Marsland (Athens: Ohio University Press, 2014), 3; Livingston, *Debility and the Moral Imagination*, 16; Neil Kodesh, *Beyond the Royal Gaze: Clanship and Public Healing in Buganda* (Charlottesville: University of Virginia Press, 2010); and Meredith Turshen, *The Politics of Public Health* (New Brunswick, New Jersey: Rutgers University Press, 1989). Public health and medical science are also fundamental to what Geissler has termed the "para-state." This move to account for the ongoing relevance of the state and various publics, even as state sovereignty particularly over heath is hindered by the prevalence of nonstate actors and organizations, is crucial to the analysis here. See Wenzel Geissler, ed., *Para-States and Medical Science: Making African Global Health* (Durham, NC: Duke University Press, 2015); Vinh-Kim Nguyen, *The Republic of Therapy: Triage and Sovereignty in West Africa's Time of AIDS* (Durham, NC: Duke University Press, 2010); Susan Reynolds Whyte, Michael A. Whyte, Lotte Meinert, and Jenipher Twebaze, "Therapeutic Clientship: Belonging in Uganda's Projectified Lanscape of AIDS Care," in Biehl and Petryna, *When People Come First*, 140–65; and Susan Reynolds Whyte, ed., *Second Chances: Surviving AIDS in Uganda*, (Durham, NC: Duke University Press, 2014). For public health more broadly, see Christopher Hamlin, "Public Health," in *The Oxford Handbook of the History of Medicine*, ed. Mark Jackson (Oxford: Oxford University Press, 2011), 411–428; Dorothy Porter, *Health, Civilization and the State: A History of Public Health from Ancient to Modern Times* (New York: Routledge, 1999); and George Rosen, *A History of Public Health*, extended ed. (Baltimore: Johns Hopkins University Press, 1993).

27. Beyond showing how science and medicine "circulated" in ways that undermine commonly held views of metropolitan importance vis-à-vis colonial and postcolonial peripheries, the history of nutrition in Uganda also echoes Clare Wendland's assertion that the notion of "Western" biomedicine as somehow not "African" is a view that mistakes a resource gap and the resulting constraints on biomedical provision for something more. Claire L. Wendland, *A Heart for the Work: Journeys through an*

African Medical School (Chicago: University of Chicago Press, 2010). For recent work on circulations see: Warwick Anderson, "Making Global Health History: The Postcolonial Worldliness of Biomedicine," *Social History of Medicine* 27, no. 2 (2014): 372–84; Hansjörg Dilger, Abdoulaye Kane, and Stacey Ann Langwick, eds., *Medicine, Mobility, and Power in Global Africa: Transnational Health and Healing* (Bloomington: Indiana University Press, 2012); and Boyd, *Preaching Prevention.*

28. Note on terminology: Buganda is the name of the kingdom, Luganda is the language spoken in Buganda, and Ganda refers to the people of Buganda, or the Baganda.

29. Richard Reid, *Political Power in Pre-Colonial Buganda: Economy, Society and Warfare in the Nineteenth Century* (Athens: Ohio University Press, 2002); David Schoenbrun, *A Green Place, A Good Place: Agrarian Change, Gender, and Social Identity in the Great Lakes Region to the 15th Century* (Portsmouth, NH: Heinemann, 1998); Holly Elisabeth Hanson, *Landed Obligation: The Practice of Power in Buganda* (Portsmouth, NH: Heinemann, 2003); Neil Kodesh, "History from the Healer's Shrine: Genre, Historical Imagination, and Early Ganda History," *Comparative Studies in Society and History* 49, no. 3 (2007): 527–52; Kodesh, *Beyond the Royal Gaze*; and Shane Doyle, *Crisis and Decline in Bunyoro: Population and Environment in Western Uganda, 1860-1955* (Athens: Ohio University Press, 2006).

30. D. A. Low and R. C. Pratt, *Buganda and British Overrule: Two Studies* (London: Oxford University Press, 1970); Michael Twaddle, "The Muslim Revolution in Buganda," *African Affairs* 71, no. 282 (1972): 54–72; D. A. Low, *Buganda in Modern History* (Berkeley: University of California Press, 1971); Michael Twaddle, *Kakungulu and the Creation of Uganda* (London: James Currey, 1993); and A. D. Roberts, "The Sub-Imperialism of the Baganda," *Journal of African History* 3 (1962): 435–50.

31. Michael Twaddle, "The *Bakungu* Chiefs of Buganda under British Colonial Rule, 1900-1930," *Journal of African History* 10 (1969): 309–22.

32. Low, *Buganda in Modern History*; Christopher Wrigley, "The Changing Economic Structure of Buganda," in *The King's Men: Leadership and Status in Buganda on the Eve of Independence*, ed. L. A. Fallers (London: Oxford University Press, 1964), 16–63; Audrey I. Richards, Ford Sturrock, and Jean M. Fortt, eds., *Subsistence to Commercial Farming in Present-Day Buganda: An Economic and Anthropological Survey* (Cambridge: Cambridge University Press, 1973); and Henry W. West, *Land Policy in Buganda* (London: Cambridge University Press, 1972).

33. Charles M. Good, *The Steamer Parish: The Rise and Fall of Missionary Medicine on an African Frontier* (Chicago: University of Chicago Press, 2004).

34. For sleeping sickness see: Maryinez Lyons, *The Colonial Disease: A Social History of Sleeping Sickness in Northern Zaire, 1900-1940* (Cambridge: Cambridge University Press, 1992); Michael Worboys, "The Comparative History of Sleeping Sickness in East and Central Africa, 1900–1914," *History of Science* 32 (1994): 89–101; Heather Bell, *Frontiers of Medicine in the Anglo-Egyptian Sudan, 1899-1940* (Oxford: Clarendon Press, 1999); Kirk Arden Hoppe, *Lords of the Fly: Sleeping Sickness Control in British East Africa, 1900-1960* (Westport, CT: Praeger, 2003); and Mari Webel, "Medical Auxiliaries and the Negotiation of Public Health in Colonial

Northwestern Tanzania," *Journal of African History* 54, no. 3 (2013): 393–416. For venereal infections and medical training in Uganda see: John Iliffe, *East African Doctors: A History of the Modern Profession* (Cambridge: Cambridge University Press, 1998); Arthur W. Williams, "The History of Mulago Hospital and the Makerere College Medical School," *East African Medical Journal* 29, no. 7 (1952): 253–63; Shane Doyle, *Before HIV: Sexuality, Fertility and Mortality in East Africa, 1900-1980* (Oxford: Oxford University Press, 2013); Michael William Tuck, "Syphilis, Sexuality, and Social Control: A History of Venereal Disease in Colonial Uganda," (Ph.D. Northwestern University, 1997); and Diane Zeller, "The Establishment of Western Medicine in Buganda," (Ph.D. thesis, Columbia University, 1971).

35. Iliffe, *East African Doctors*; Carol Sicherman, *Becoming an African University: Makerere, 1922-2000* (Trenton, NJ: Africa World Press, 2005); and Alexander Mwa Odonga, *The First Fifty Years of Makerere Medical School: And the Foundation of Scientific Medical Education in East Africa* (Kisubi, Uga.: Marianum Press, 1989).

36. This use of the term *medicalization* builds upon, but differs somewhat from, a longstanding emphasis (primarily in sociology and medical anthropology) on a process in which components of life that were not previously deemed pathological came to be defined as medical problems. The insights that emerged from an analysis of how social conditions and aspects of everyday life became medical concerns were critical to an understanding of medical science as a social and cultural undertaking. Several scholars posit that medicalization has recently taken on novel forms (biomedicalization and pharmaceuticalization) as further technoscientific innovations have extended and transformed the process itself. Vinh-Kim Nguyen argues that the concept of medicalization is ill-suited to at least many African contexts (and to critiques of treatment as prevention) because it "presumes that there is a social sphere to medicalize," whereas in many cases this may not be the case, at least not in the same way that the theory of medicalization posits. In this study, the central distinction is not between the social and the biomedical, but between curative and preventive medicine, or between clinical medicine and the promotion of public health. Vinh-Kim Nguyen, "Treating to Prevent HIV: Population Trials and Experimental Societies," in Geissler, *Para-States and Medical Science*, 47–77. See Adele E. Clarke et al., "Biomedicalization: Technoscientific Transformations of Health, Illness, and U.S. Biomedicine," *American Sociological Review* 68, no. 2 (2003): 161–94; and Antonio Maturo, "Medicalization: Current Concept and Future Directions in a Bionic Society," *Mens Sana Monographs* 10, no. 1 (2012): 122–33. For pharmaceuticalization see: João Guilherme Biehl, *Will to Live: Aids Therapies and the Politics of Survival* (Princeton: Princeton University Press, 2007). For a more recent discussion of medicalization see: Randall Packard, *A History of Global Health: Interventions into the Lives of Other Peoples* (Baltimore, MD: Johns Hopkins University Press, 2016).

Chapter 1: Diagnostic Uncertainty and Its Consequences

1. Hunt, *A Nervous State*.

2. The shared symptoms include a skin rash, enlarged liver, irritability, and an overall failure to thrive. The differential diagnosis was particularly difficult in the

context of early twentieth-century biomedical knowledge and in the absence of clinical screening.

3. Tuck, "Syphilis, Sexuality, and Social Control," 41–47; and J. N. P. Davies, "The History of Syphilis in Uganda," *Bulletin of the World Health Organization* 15, no. 6 (1956): 1041–55.

4. Albert Ruskin Cook, *Uganda Memories (1897–1940)* (Kampala: Uganda Society, 1945), 244. Figures cited in Cook's account report 8,572 births and 15,011 deaths; Cyril Ehrlich, "The Uganda Economy, 1903–1945," in *History of East Africa*, ed. Vincent Harlow and E. M. Chilver (Oxford: Oxford University Press, 1965), 395–475; Megan Vaughan, *Curing their Ills: Colonial Power and African Illness* (Stanford: Stanford University Press, 1991); Megan Vaughan, "Syphilis in Colonial East Africa: The Social Construction of an Epidemic," in *Epidemics and Ideas: Essays on the Historical Perception of Pestilence*, ed. T. Ranger and P. Slack (New York: Cambridge University Press, 1992), 269–302; Carol Summers, "Intimate Colonialism: The Imperial Production of Reproduction in Uganda, 1907-1925," *Signs: Journal of Women in Culture and Society* 16, no. 4 (1991): 787–807; M. W. Tuck, "Venereal Disease, Sexuality and Society in Uganda" in *Sex, Sin, and Suffering: Venereal Disease and European Society since 1870*, ed. R. Davidson and L. Hall (New York: Routledge, 2001), 191–203; and Nakanyike Musisi, "The Politics of Perception or Perception as Politics? Colonial and Missionary Representations of Baganda Women, 1900–1945," in *Women in African Colonial Histories*, ed. Jean Allman, Susan Geiger, and Nkanyike Musisi (Bloomington: Indiana University Press, 2002), 95–115.

5. Iliffe, *East African Doctors*; Sicherman, *Becoming an African University*; Odonga, *First Fifty Years*; and Williams, "The History of Mulago Hospital."

6. Doyle, *Before HIV*, 87–95; and Tuck, "Syphilis, Sexuality, and Social Control."

7. "Report of the Lady Coryndon Maternity Training School, Namirembe, 1931," M/7 A7/4 Church Missionary Society Archive, Cadbury Research Library, University of Birmingham (hereafter CMS); "Report of the Lady Coryndon Maternity Training School, Namirembe, 1932," M/7 A7/5 CMS; Meeting of the Lady Coryndon M.T.S. Committee Minutes, January 2, 1934, M/Y A7/7 CMS; and Correspondence from R. Y. Stones, n.d. M/Y A7/10 CMS, Uganda Mission Correspondence, 1938–1939.

8. Balance Sheet: Mengo Hospital, Uganda, 1931 M/Y A7/5 CMS; and Tuck, "Syphilis, Sexuality, and Social Control," 217–26.

9. Hugh Trowell interviewed by Elizabeth Bray, MSS.Afr.s.1872 (144B), Rhodes House Library, Oxford University (hereafter RHL), 15–16.

10. H. C. Trowell, "Pellagra in African Children," *Archives of Disease in Childhood* 12 (1937): 208.

11. Trowell, "Pellagra in African Children," 205.

12. Trowell, interview, MSS.Afr.s.1872 (144B), RHL, 17; and H. C. Trowell, "Food, Protein and Kwashiorkor: A Presidential Address, 1956," *Uganda Journal* 21 (1957): 84–85.

13. Trowell, "Food, Protein and Kwashiorkor," 84–85.

14. H. S. Stannus, "A Nutritional Disease of Childhood Associated with a Maize Diet—and Pellagra," *Archives of Disease in Childhood* 9, no. 50 (1934): 115–18. For more on Hugh Stannus and his work in Nyasaland, see Vaughan, *Curing Their Ills*, especially chapter 2.

15. C. D. Williams, "Kwashiorkor: A Nutritional Disease of Children Associated with a Maize Diet," *Lancet*, November 16, 1935, 1151–52.

16. Celia Petty, "Primary Research and Public Health: The Prioritization of Nutrition Research in Inter-war Britain," in *Historical Perspectives on the Role of the MRC*, ed. J. Austoker and L. Bryder (New York: Oxford University Press, 1989), 83–108; and E. V. McCollum, *The Newer Knowledge of Nutrition* (New York: The MacMillan Company, 1918).

17. Roger Whitehead, "Kwashiorkor in Uganda," in *The Contribution of Nutrition to Human and Animal Health*, ed. E. M. Widdowson and J. C. Mathers (Cambridge: Cambridge University Press, 1992), 305.

18. Williams became the first Head of Maternal and Child Health at the World Health Organization (WHO).

19. Trowell, interview, MSS.Afr.s.1872 (144B), RHL, 20–21.

20. Trowell, "Pellagra in African Children," 193–212; and H. C. Trowell, "A Note on Infantile Pellagra," *Transactions of the Royal Society of Medicine and Hygiene* 35, no. 1 (1941): 18–20.

21. Trowell, interview, MSS.Afr.s.1872 (144B), RHL, 32.

22. "Death of a Native at a Mission Hospital," *East African Standard* (May 6, 1933), M/Y A7/6 CMS; "Dear Stones," May 17, 1938, M/Y A7/10 CMS; "Resume of Conversations Between Dr. Kauntze and Dr. Stones," March 17, 1938, M/Y A7/10 CMS; and "Medical Sub-Conference Minutes," August 17, 1939, M/Y A7/10 CMS.

23. Cook, *Uganda Memories (1897–1940)*; Albert R. Cook, "The Medical History of Uganda: Part II," *East African Medical Journal*, 13 (1937): 99–110; interview with Dr. Paget Stanfield, Edinburgh, Scotland, November 27, 2003; and "Dr. Hebe Flower Welbourn: Reminiscences of My Career in Uganda," Personal Memoir, MSS.Afr.s.1872 (152), RHL, 5.

24. Trowell, interview, MSS.Afr.s.1872 (144B), RHL, 21; and Trowell, "Food, Protein and Kwashiorkor," 85.

25. H. C. Trowell, J. N. P. Davies, and R. F. A. Dean, *Kwashiorkor* (1954; reprint, London: Academic Press, 1982), 179. The provision of nicotinic acid was later viewed as potentially dangerous because it meant the body used compounds that would otherwise be available for "more vital" processes of synthesis and recovery.

26. Trowell, interview, MSS.Afr.s.1872 (144B), RHL, 27–29.

27. Trowell, interview, MSS.Afr.s.1872 (144B), RHL, 35–37; and Iliffe, *East African Doctors*, 87.

28. E. M. K. Muwazi, H. C. Trowell, and J. N. P. Davies, "Congenital Syphilis in Uganda," *East African Medical Journal* 24, no. 4 (April 1947): 152–70; H. C. Trowell and E. M. K. Muwazi, "A Contribution to the Study of Malnutrition in Central Africa: A Syndrome of Malignant Malnutrition," *Transactions of the Royal Society of*

Tropical Medicine and Hygiene 39, no. 3 (1945): 229–43; and H. C. Trowell and E. M. K. Muwazi, "Severe and Prolonged Underfeeding in African Children (The Kwashiorkor Syndrome of Malignant Malnutrition)," *Archives of Disease in Childhood* 20 (1945): 110–16.

29. Muwazi, Trowell, and Davies, "Congenital Syphilis in Uganda," 156 and 162.

30. Trowell, interview, MSS.Afr.s.1872 (144B), RHL, 33.

31. Trowell and Muwazi, "Severe and Prolonged Underfeeding"; and Trowell and Muwazi, "Study of Malnutrition."

32. Muwazi, Trowell, and Davies, "Congenital Syphilis in Uganda,"154.

33. Trowell, interview, MSS.Afr.s.1872 (144B), RHL, 37.

34. Trowell, Davies, and Dean, *Kwashiorkor*, devotes an entire section to "Protein Malnutrition in Adults," 249–82.

35. Trowell, interview, MSS.Afr.s.1872 (144B), RHL, 42.

36. Audrey Richards, ed., *Economic Development and Tribal Change: A Study of Immigrant Labour in Buganda* (Cambridge: W. Heffer & Sons Ltd., 1952); and Trowell and Muwazi, "Study of Malnutrition," 230–31.

37. Trowell and Muwazi, "Study of Malnutrition," 230 and 234.

38. Muwazi, Trowell, and Davies, "Congenital Syphilis in Uganda," 154.

39. M. Stanier and M. D. Thompson, "The Serum Protein Levels of Newborn African Infants," *Archives of Disease in Childhood* 29, no. 144 (1954): 110–12. The precise dates in which a study was conducted are often difficult to determine. Scientists take great care to accurately detail the methods employed and the number of cases included in a study, but the date of the actual investigation is only in very exceptional cases mentioned in the scientific articles and reports. It is possible to establish likely dates for many studies based upon the date of publication, dates cited by the authors, and knowledge of the overall trajectory of the scientific findings and investigations, but such dates remain speculative. This case is a particular anomaly as it was published in 1954 but was representative of the work conducted in the late 1940s. Several factors may have delayed its publication, but further investigation would be required to establish when the blood was in fact extracted for the study.

40. Holmes to Lewthwaite, March 26, 1951, FD 1/1873, Public Record Office (hereafter PRO); "Dr. Himsworth's Visit to Africa," January 1951, FD 1/1869, PRO; and Trowell, interview, MSS.Afr.s.1872 (144B), RHL, 42.

41. E. G. Holmes et al., "Red Blood Count of East African Students," *East African Medical Journal* 27, no. 9 (1950): 360–70; E. G. Holmes et al., "An Investigation of Serum Proteins of Africans in Uganda," *Transactions of the Royal Society of Tropical Medicine and Hygiene* 45, no. 3 (1951): 371; and Holmes to Lewthwaite, FD 1/1873 PRO. It is, again, difficult to determine when these two studies were conducted (see note 39 above). Despite the lack of definitive evidence, it is highly likely that they were inaugurated and perhaps even completed prior to the insurrection. It is clear that from at least 1947 onward, it became possible to conduct serum protein fractionations at the physiology and biochemistry lab in Uganda.

42. Interview with Professor C. M. Ndugwa, Mulago Hospital, Kampala, Uganda, June 8, 2004. For more on how upheaval at Budo was central to political activism in this period, see Carol Summers, "'Subterranean Evil' and 'Tumultuous Riot' in Buganda: Authority and Alienation at King's College, Budo, 1942," *Journal of African History* 47, no. 1 (2006): 93–113.

43. Trowell, interview, MSS.Afr.s.1872 (144B), RHL, 53.

44. Trowell, interview, MSS.Afr.s.1872 (144B) RHL; "Dr. Hugh Trowell OBE, MD, FRCP," Personal Memoir, MSS.Afr.s.1872 (144A), RHL; and "Davies, Dr. J.N.P" Personal Memoir, MSS.Afr.s.1872 (40), RHL.

45. J. C. Waterlow, H. C. Trowell, and J. N. P. Davies, "Liver Biopsy Demonstrations," in *Malnutrition in African Mothers, Infants, and Young Children: Report of the Second Inter-African Conference on Nutrition, Fajara, Gambia, 19–27 November, 1952* (London: Her Majesty's Stationary Office, 1954), 100.

46. E. G. Holmes, R. E. Jones, M. W. Stanier, "The Protein Metabolism of East Africans with Hookworm Anaemia and Other Conditions," FD 1/1873, PRO. This study includes evidence that predates the insurrection and evaluations are recorded from 1948 and 1949. Moreover, the study was undertaken after earlier nitrogen balance studies yielded remarkable results that required follow-up research. The large majority of the estimations cited in this report are from 1950 and 1951.

47. Ibid., 3.

48. Trowell, interview, MSS.Afr.s.1872 (144B) RHL.

49. Trowell, interview, MSS.Afr.s.1872 (144B), RHL, 36; "Dr. Hugh Trowell," MSS.Afr.s.1872 (144A), RHL,17; and "Davies, Dr. J.N.P," MSS.Afr.s.1872 (40), RHL.

50. Trowell, interview, MSS.Afr.s.1872 (144B), 36.

51. J. N. P. Davies, "The Essential Pathology of Kwashiorkor," *Lancet*, 1948, 317–20.

52. "Davies, Dr. J.N.P," MSS.Afr.s.1872 (40), RHL, 74.

53. Ibid., 75.

54. Ibid.

55. "Joint FAO/WHO Expert Committee on Nutrition: Report on the First Session," World Health Organization Technical Report Series No. 16 (Geneva: WHO, 1950), 15; and Trowell, interview, MSS.Afr.s.1872 (144B), RHL, 56–57.

56. J. F. Brock and M. Autret, *Kwashiorkor in Africa*, World Health Organization: Monograph Series 8 (Geneva: WHO, 1952).

57. "Joint FAO/WHO Expert Committee on Nutrition: Report on the Second Session," World Health Organization Technical Report Series No. 44 (Geneva: WHO, 1951), 29.

58. M. Autret and M. Behar, *Sindrome Policarencial Infantil (Kwashiorkor) and its Prevention in Central America*, FAO Nutritional Studies 13 (Rome: FAO, 1954); and J. C. Waterlow and A. Vergara, *Protein Malnutrition in Brazil*, FAO Nutritional Studies 14 (Rome: FAO, 1956).

59. "Davies, Dr. J.N.P," MSS.Afr.s.1872(40), RHL, 76.

60. Trowell, interview, MSS.Afr.s.1872 (144B), RHL, 43.

61. "Uganda Protectorate, Civil Disturbances, 1949: Memorandum by Resident, Buganda," CO 537/4679, PRO; Low, "The Advent of Populism in Buganda," in *Buganda in Modern History*, 139–65; Carol Summers, "Radical Rudeness: Ugandan Social Critiques in the 1940s," *Journal of Social History* 39, no. 3 (2006): 741–70; Summers, "Young Buganda and Old Boys: Youth, Generational Transition, and Ideas of Leadership in Buganda, 1920–1949," *Africa Today* 51, no. 3 (2005): 120–21; Summers, "Grandfathers, Grandsons, Morality, and Radical Politics in Late Colonial Buganda," *International Journal of African Historical Studies* 38, no. 3 (2005): 427–47; and Summers, "Young Africa and Radical Visions: Revisiting the Bataka in Buganda, 1944–54," unpublished paper, March 2004.

62. According to Summers, the targets of the insurrection were not randomly selected and British officials were explicitly not attacked. She notes that "activists shouted at each other to remember the Agreement and not aim at the Britons" (Summers, "Young Africa and Radical Visions," n75). Instead, activists sought to destroy "houses owned (or rumoured to be owned) by chiefs considered close to British administration" (Summers, "Radical Rudeness," 751).

63. Black to Director of Medical Services, 16 May 1949, Uganda Ministry of Health (UMOH) 1/A, quoted in Iliffe, *East African Doctors*, 91.

64. Ladkin to Hennessey, 16 May 1949, UMOH 1/A, quoted in Iliffe, *East African Doctors*, 91.

65. Trowell, interview, MSS.Afr.s.1872 (144B), RHL, 43.

66. Trowell, "Food, Protein and Kwashiorkor," 84; Trowell, interview, MSS. Afr.s.1872 (144B), RHL, 33–34; and Brock and Autret, *Kwashiorkor in Africa*, 24.

67. Ladkin to Hennessey, 16 May 1949, UMOH 1/A, quoted in Iliffe, *East African Doctors*, 91.

68. Gardner Thompson, "Colonialism in Crisis: The Uganda Disturbances of 1945," *African Affairs* 91 (1992): 607.

69. For one example from this period: "G.R. Kizza of Kalokwe, Gomboloa of Musule, Busiro, in Gambuze, 26 April 1946," in *The Mind of Buganda: Documents of the Modern History of an African Kingdom*, ed. D. A. Low (Berkeley: University of California Press, 1971), 131–33.

70. Low, "Advent of Populism"; and Summers, "Grandfathers, Grandsons, Morality," 427–47.

71. Summers, "Grandfathers, Grandsons, Morality," 428.

72. "Report of an audience by Kabaka Mutesa II to 'eight peoples' representatives', 25 April 1949," in Low, *Mind of Buganda*, 138–39. Caution must be exercised in associating these references to malnutrition with kwashiorkor, as consensus within the biomedical community was only beginning to grow in favor of the protein theory by 1949 and the evidence does not suggest that, in this period, local conceptions of the condition featured malnutrition as a primary cause.

73. Iliffe, *East African Doctors*, 84.

74. L. A. Fallers and S. B. K. Musoke, "Social Mobility, Traditional and Modern" in Fallers, *King's Men*, 193.

75. Interview with Dr. Luis Mugambe Muwazi, Kampala, Uganda, April 2004. Muwazi's formal political career was associated with the Kabaka Yekka (Kabaka Alone) party that emerged as the focus of popular politics in Buganda in the context of the exile of Kabaka Mutesa II in 1953.

76. A now extensive scholarship exists on the topic of blood-taking "rumors," especially in East Africa, including especially: Luise White, *Speaking with Vampires: Rumor and History in Colonial Africa* (Berkeley: University of California Press, 2000); White, "'They Could Make Their Victims Dull': Genders and Genres, Fantasies and Cures in Colonial Southern Uganda," *The American Historical Review* 100, no. 5 (1995), 1379–1402; P. Wenzel Geissler, "'Kachinja Are Coming!': Encounters around Medical Research Work in a Kenyan Village," *Journal of the International African Institute* 75, no. 2 (2005): 173–202; James Fairhead, Melissa Leach, and Mary Small, "Where Techno-Science Meets Poverty: Medical Research and the Economy of Blood in The Gambia, West Africa," *Social Science and Medicine* 63 (2006): 1109–20; P. Wenzel Geissler, Ann Kelly, Babatunde Imoukhuede, and Robert Pool, "'He Is Now Like a Brother, I Can Even Give Him Some Blood'—Relational Ethics and Material Exchanges in A Malaria Vaccine 'Trial Community' in The Gambia," *Social Science and Medicine* 67 (2008): 696–707; Melissa Graboyes, "Fines, Orders, Fear . . . and Consent?: Medical Research in East Africa, c. 1950s," *Developing World Bioethics* 10, no. 1 (2010): 34–41; and Graboyes, *The Experiment Must Continue.*

77. Arthur Williams to Harold Himsworth, December 5, 1950, "East Africa, Medical Research In," FD 1/1869, PRO.

78. Trowell, interview, MSS.Afr.s.1872 (144B), RHL, 57.

79. Ibid., 52.

80. Ibid., 43.

81. Ibid., 53.

82. "Himsworth's Visit to Africa," FD 1/1869, PRO.

83. Thompson to Himsworth, January 5, 1955, FD 1/8893, PRO.

84. Hebe F. Welbourn, "Child Welfare in Mengo District, Uganda," *Journal of Tropical Pediatrics*, June 1956, 28–29.

85. Whitehead, interview.

86. H. J. L. Burgess, "The Medical Research Council's Rural Child Welfare Clinic," *East African Medical Journal*, May 1960, 391–98.

87. R. F. A. Dean, "The Treatment of Kwashiorkor with Milk and Vegetable Proteins," *British Medical Journal* 2, no. 4788 (1952): 792; and "Tropical Medical Research Board, Progress Report 1955-61 of the Infantile Malnutrition Research Unit, Mulago Hospital, Kampala, Uganda," and "Infantile Malnutrition Research Unit, Office Note" December 7, 1961, FD 12/273, PRO.

88. Dean, "Treatment of Kwashiorkor," 792. Dr. Margaret Thompson also cites achieving mortality rates of between 10 and 18 percent in this period: Thompson to Himsworth, January 25, 1954, and "Protein Shortage in African Children: Some Causes and Effects," January 1953, FD 1/8893, PRO. For more on olumbe, see M. Southwold, "Ganda Conceptions of Health and Disease," in *Attitudes to Health and*

Disease Among Some East African Tribes (Kampala: East African Institute of Social Research, 1959), 44.

89. R. G. Whitehead and R. F. A. Dean, "Serum Amino Acids in Kwashiorkor: I. Relationship to Clinical Condition," *American Journal of Clinical Nutrition* 14, no. 6 (1964): 313–14; and R. G. Whitehead, "The Assessment of Nutritional Status in Protein-Malnourished Children," *Proceedings of the Nutrition Society* 28 (1969): 2.

90. Ruth Scwhartz and R. F. A. Dean, "The Serum Lipids in Kwashiorkor: I. Neutral Fat, Phospholipids and Cholesterol," *Journal of Tropical Pediatrics* 3, no. 1 (June 1957): 25.

91. R. F. A. Dean and Ruth Schwartz, "The Serum Chemistry in Uncomplicated Kwashiorkor," *British Journal of Nutrition* 7 (1953): 134.

92. Trowell, Davies, and Dean, *Kwashiorkor*, 215.

93. R. G. Whitehead and T. R. Milburn, "Amino Acid Metabolism in Kwashiorkor, II. Metabolism of Phenylalanine and Tyrosine" *Clinical Science* 26 (1964): 279; and R. G. Whitehead, "Amino Acid Metabolism in Kwashiorkor, I. Metabolism of Histidine and Imidazole Derivatives," *Clinical Science* 26 (1964): 271–78.

94. Mary K. Clegg and R. F. A. Dean, "Balance Studies on Peanut Biscuit in the Treatment of Kwashiorkor," *American Journal of Clinical Nutrition* 8 (1960): 886–87.

95. See, for example, C. W. Hattersley, *The Baganda at Home: With One Hundred Pictures of Life and Work in Uganda* (London: The Religious Tract Society, 1908), 223–25; Aidan Southall and Peter Gutkind, in their investigation of urban life in the mid-1950s, found that "Although a great many people come to Mulago and reside there so that they may avail themselves of the services provided by the Hospital . . . even with such matters as a broken arm or an infection, considerable numbers of people, both immigrant and Ganda, educated and uneducated, visit local African 'doctors' first." Southall and Gutkind, *Townsmen in the Making: Kampala and Its Suburbs* (Kampala: East African Institute of Social Research, 1956), 125.

96. Bennett, "V. Concepts of Disease," 148.

97. Southwold, "Ganda Conceptions of Health," 44.

98. Southwold, "Ganda Conceptions of Health," 44; Southall and Gutkind, *Townsmen in the Making*, 125; and Bennett, "V. Concepts of Disease," 153.

99. Marc H. Dawson, "The 1920s Anti-Yaws Campaigns and Colonial Medical Policy in Kenya," *International Journal of African Historical Studies* 20, no. 3 (1987): 423–28.

100. L. P. Mair, *An African People in the Twentieth Century* (1934; reprint, New York: Russell & Russell, 1965), 259. (Emphasis in original).

101. "Welbourn: Reminiscences," MSS.Afr.s.1872 (152), RHL, 5.

102. This interpretation differs slightly from Jennifer Tappan, "Blood Work and Rumors of Blood: Nutritional Research and Insurrection in Buganda, 1938–1952," *International Journal of African Historical Studies* special edition on "Incorporating Medical Research into the History of Medicine in East Africa," 47, no. 3 (2014): 473–94.

103. M. D. Thompson, "Second Interim Report to the Medical Research Council: The Etiology, Prevention and Treatment of the Essential Protein

Deficiency of Kwashiorkor: A. The Type of Protein Required and The Possibility of a Local Vegetable Substitute for Milk," adjoining correspondence: Thompson to Himsworth, FD 1/8893, PRO, 6.

104. Thompson, "Protein Shortage," FD 1/8893, PRO, 2.

105. Ibid.

106. Trowell and Muwazi, "Severe and Prolonged Underfeeding," 114.

107. Thompson, "Protein Shortage," FD 1/8893 PRO, 6.

108. Thompson and Trowell, "Pancreatic Enzyme Activity," 1033.

109. Latimer K. Musoke, "An Analysis of Admissions to the Paediatric Division, Mulago Hospital in 1959," *Archives of Disease in Childhood* 36 (1961): 305.

110. Trowell, Davies, and Dean, *Kwashiorkor*, 63.

111. M. A. Church, "Evaluation as an Integral Aspect of Nutrition Education," September 1982, Personal Papers of Dr. Mike Church, 7; A. L. Kitching and G. R. Blackledge, *A Luganda-English and English-Luganda Dictionary* (London: Society for Promoting Christian Knowledge, 1925). In Luganda: *akaana kalina obwosi*, meaning "the very small child or baby has or is suffering from premature weaning" as the term obwosi is a derivative of the verb *kwosera* or "to wean too soon." R. A. Snoxall, ed., *Luganda-English Dictionary* (Oxford: Clarendon Press, 1967). The concept of *obwosi* appears to be exceptional in this sense, as most "other diseases are said to 'bite' one; e.g. 'Musujja gunnuma', 'Fever bites me'; but with obwosi one says that the child has obwosi" (Southwold, "Ganda Conceptions of Health and Disease," 46) (Emphasis in original). For more on categories of illness that, like obwosi, were associated with particular situations rather than specific symptoms, see Bennett, "V. Concepts of Disease," 157.

112. F. J. Bennett and J. S. W. Lutwama, "Buganda Women: What They Want to Know," *International Journal of Health Education* 5, no. 2 (1962): 80.

113. Bennett, "V. Concepts of Disease," 153; and Hebe F. Welbourn, "Weaning among the Baganda," *Journal of Tropical Pediatrics and African Child Health* 9 (June 1963): 20.

114. Maurice King, ed., *Medical Care in Developing Countries: A Primer on the Medicine of Poverty and a Symposium from Makerere* (Nairobi, Ken.: Oxford University Press, 1966), 14:17g; Bennett, "V. Concepts of Disease," 152–55. The hypothermia that accompanied severe acute malnutrition was seen as linked to cold feet and cold air of empewo and ekiglanga, respectively (Church, "Nutrition Education," 7).

115. Bennett, "V. Concepts of Disease," 152.

116. Mary D. Salter Ainsworth, *Infancy in Uganda: Infant Care and the Growth of Love* (Baltimore: John's Hopkins Press, 1967), 115.

117. Wenzel P. Geissler, and Catherine Molyneux, *Evidence, Ethos and Experiment: The Anthropology and History of Medical Research in Africa* (New York: Berghahn, 2011).

118. This point has most recently and perhaps most poignantly been observed in the 2014 West African Ebola epidemic, in which over eleven thousand people lost their lives.

1. Whitehead, "Kwashiorkor in Uganda," 307. In this period, mortality rates ranged from 40 to 60 percent in Uganda and up to 90 percent in Africa as a whole (see chapter 1).

2. R. F. A. Dean, "Preface" in *Plant Proteins in Child Feeding*, Medical Research Council Special Report Series 279 (London: H. M. Stationery Office, 1953); and Department of Experimental Medicine, Cambridge, and Associated Workers, *Studies of Undernutrition, Wuppertal 1946–9*, Medical Research Council Special Report Series 275 (London: H. M. Stationery Office, 1951).

3. Whitehead, "Kwashiorkor in Uganda," 307; and Whitehead, interview.

4. Dean, "Treatment of Kwashiorkor," 792; R. F. A. Dean and B. Weinbren, "Fat Absorption in Chronic Severe Malnutrition in Children," *Lancet* 268, no. 6936 (August 4, 1956): 252; R.F.A. Dean and M. Skinner, "A Note on the Treatment of Kwashiorkor," *Journal of Tropical Pediatrics*, March 1957, 215–16; Whitehead to Himsworth, November 21, 1964, FD 12/274, PRO; Stanfield, interview; and Whitehead, interview.

5. Whitehead, "Kwashiorkor in Uganda," 307; and "MRC Child Nutrition Unit: A Progress Report from October 1968–December 1970 Presented to the Tropical Medical Research Board of the Medical Research Council of the United Kingdom and the National Research Council of Uganda," FD 12/282 PRO.

6. R. F. A. Dean and D. B. Jelliffe, "Diagnosis and Treatment of Protein-Calorie Malnutrition," *Courrier* 10, no. 7 (1960): 433.

7. Ibid.; and M. A. Church, "The Hospital Treatment of Severe Kwashiorkor," January 1967, Personal Papers of Dr. Mike Church.

8. Dean, "Treatment of Kwashiorkor," 792.

9. Dean and Schwartz, "Uncomplicated Kwashiorkor," 135–36.

10. I. Schneideman, F. J. Bennett, and I. H. E. Rutishauser, "The Nutrition Rehabilitation Unit at Mulago Hospital-Kampala: Development and Evaluation, 1965–67," *Journal of Tropical Pediatrics* 17, no. 1 (1971): 26.

11. Trowell, Davies, and Dean, *Kwashiorkor*, 206–7.

12. For more on the importance of a child's appetite and acceptance of the formula, see Barbara Cooper, "Chronic Malnutrition and the Trope of the Bad Mother," in *A Not-So Natural Disaster: Niger 2005*, ed. Xavier Crombé and Jean-Hervé Jézéquel (New York: Columbia University Press, 2009), 147–68.

13. Dean, "Treatment of Kwashiorkor," 792–93.

14. Church, "Nutrition Education," 4.

15. Interview with Drs. Mike Church and Paget Stanfield, Edinburgh, Scotland, November 26, 2003; and Church, "Nutrition Education," 4.

16. J. F. Brock, "Dietary Proteins in Relation to Man's Health," *Federation Proceedings* 20, no. 1 (1961): 61; and K. J. Carpenter, *Protein and Energy: A Study of Changing Ideas in Nutrition* (New York: Cambridge University Press, 1994), 160.

17. National Institutes of Health and Committee on Protein Malnutrition, Food and Nutrition Board, *Progress in Meeting Protein Needs of Infants and Preschool*

Children, Proceedings on an International Conference held in Washington, DC, August 21–24, 1960, 541.

18. "Report on the Second Session," 22–27; "Joint FAO/WHO Expert Committee on Nutrition: Report on the Third Session," World Health Organization Technical Report Series No. 72 (Geneva: WHO, 1953). In addition to the regular meetings of the Joint FAO/WHO Expert Committee on Nutrition, see: J. C. Waterlow, ed., *Protein Malnutrition, Proceedings of a Conference in Jamaica, Sponsored Jointly by the FAO, WHO and the Josiah Macy Jr. Foundation, New York* (Cambridge: Cambridge University Press, 1953); Uganda Government, WHO, and FAO, *Nutrition Seminar for English-Speaking Countries and Territories in Africa South of the Sahara*, FAO Report no. 960 (Rome: FAO, WHO, 1958); FAO and WHO, *Report of the FAO/WHO Seminar on Problems of Food and Nutrition in Africa South of the Sahara*, FAO Nutrition Meeting Report Series No. 25 (Rome: N.p., 1961); and "Progress in Meeting Protein Needs of Infants and Preschool Children," Proceedings on an International Conference held in Washington, DC, August 21–24, 1960.

19. Adults were frequently included in Trowell's early kwashiorkor research and an entire section ("Part IV: Protein Malnutrition in Adults," 250–67) is devoted specifically to adult cases in Trowell, Davies, and Dean, *Kwashiorkor.*

20. Brock and Autret, *Kwashiorkor in Africa*, 12.

21. With the exception of pregnancy, lactation, and when recovering from illness.

22. Based on a mistranslation of the term kwashiorkor as "red-boy," the 1952 report also initially proposed to confine the term kwashiorkor to the condition in Africa. This decision also reflected the emphasis placed upon the skin and hair changes that are far more pronounced in African children and children with darker complexions. As kwashiorkor quickly became a global problem of the "developing" and "Third-World" nations, the importance of ideas about biological race to conceptions of the condition became less pronounced. See Brock and Autret, *Kwashiorkor in Africa*, 12–13.

23. H. F. Welbourn, "The Danger Period during Weaning," *Journal of Tropical Pediatrics*, June 1955, 34–46; D. B. Jelliffe and R. F. A. Dean, "Protein-Calorie Malnutrition in Early Childhood (Practical Notes)," *Journal of Tropical Pediatrics*, December 1959, 96; and R. F. A. Dean, "Biochemical Changes Caused by Protein Deficiency in Young Children," *Clinica Chimica Acta* 5 (1960): 186.

24. H. F. Welbourn, "Custom and Child Health in Buganda: III. Pregnancy and Childbirth," *Tropical and Geographical Medicine* 15 (1963): 126.

25. Marcelle Geber, "The Psycho-Motor Development of African Children in the First Year, and the Influence of Maternal Behavior," *Journal of Social Psychology* 47 (1958): 194.

26. Marcelle Geber and R. F. A. Dean, "Gesell Tests on African Children," *Pediatrics*, December 1957, 1062.

27. Marcelle Geber and R. F. A. Dean, "The Psychological Changes Accompanying Kwashiorkor." *Courrier* 6, no. 1 (January 1956): 10.

28. I. H. E. Rutishauser, "Custom and Child Health in Buganda: IV. Food and Nutrition," *Tropical and Geographical Medicine* 15 (1963): 142 and 145.

29. Ibid., 145.

30. H. F. Welbourn, "The Danger Period during Weaning (Part II)," *Journal of Tropical Pediatrics*, September 1955, 102.

31. Ibid.

32. Trowell, Davies, and Dean, *Kwashiorkor*, 59.

33. Rutishauser, "IV. Food and Nutrition," 140. (Emphasis in original).

34. Dean to Williams, June 19, 1952, PP/CDW L.1, Wellcome Library (hereafter WL); R. F. A. Dean, "Observations on African Children," PP/CDW L.1, WL.

35. FAO/WHO/UNICEF Protein Advisory Group, *Lives in Peril: Protein and the Child*, World Food Problems no. 12 (Rome: N.p., 1970): 51.

36. Ibid., 15.

37. Ibid., 25.

38. Ibid., 41.

39. Brock and Autret, *Kwashiorkor in Africa*, 32–33. Dean to Williams, June 19, 1952, PP/CDW L.1, WL; R. F. A. Dean, "Observations on African Children," PP/CDW L.1, WL, 8; Welbourn, "Danger Period," 35; Davies, "Essential Pathology," 320; "Davies, Dr. J.N.P," MSS.Afr.s.1872 (40), RHL; and Trowell, interview, MSS. Afr.s.1872 (144B), RHL, 41.

40. FAO/WHO/UNICEF Protein Advisory Group, *Lives in Peril*, 41.

41. M. Autret, preface in *Encouraging the Use of Protein-Rich Foods*, by John Fridthjof (Rome: FAO, 1962), v; Carpenter, *Protein and Energy*, 160; and FAO/WHO/UNICEF Protein Advisory Group, *Lives in Peril*, 52.

42. FAO/WHO/UNICEF Protein Advisory Group, *Lives in Peril*, 25.

43. Brock and Autret, *Kwashiorkor in Africa*, 69–70.

44. "Report on the Second Session," 22–27; and "Report on the Third Session," 11–15.

45. The distinction is important as protein-rich foods need not resemble milk in regions where milk was not a "traditional" weaning food ("Report on the Third Session," 14–15).

46. "Joint FAO/WHO Expert Committee on Nutrition: Fifth Report," World Health Organization Technical Report Series No. 149 (Geneva: WHO, 1958), 19–25; and "Joint FAO/WHO Expert Committee on Nutrition: Sixth Report," World Health Organization Technical Report Series No. 245 (Geneva: WHO, 1962), 55–61.

47. Carpenter, *Protein and Energy*, 178.

48. R. F. A. Dean, "Use of Processed Plant Proteins as Human Food," in *Processed Plant Protein Foodstuffs*, ed. Aaron M. Altschul (New York: Academic Press, 1958), 205–47; and "Fifth Report," 8–9.

49. Dean to Williams, PP/CDW L.1, WL; Dean, "Treatment of Kwashiorkor," 795; and interviews with Ephraim Musoke, Luteete, Uganda, April 2004 and June 3, 2004.

50. Dean, "Treatment of Kwashiorkor," 795.

51. Musoke, interviews.

52. K. Mary Clegg, "The Availability of Lysine in Groundnut Biscuits Used in the Treatment of Kwashiorkor," *British Journal of Nutrition* 14 (1960): 325–29; and Clegg and Dean, "Balance Studies," 885–89.

53. Dean, "Treatment of Kwashiorkor," 675.

54. Whitehead, interview.

55. "Sixth Report," 56.

56. Ibid., 40.

57. FAO/WHO/UNICEF Protein Advisory Group, *Lives in Peril*, 33–39.

58. Whitehead, interview.

59. Carpenter, *Protein and Energy*, 178–179.

60. Nevin S. Scrimshaw, "A Look at the Incaparina Experience in Guatemala: The Background and History of Incaparina," *Food and Nutrition Bulletin* 2, no. 2 (1980): 1; Carpenter, *Protein and Energy*, 175. For more on "ready-to-use therapeutic food," see chapter 5.

61. Carpenter, *Protein and Energy*, 164.

62. R. F. A. Dean, "Kwashiorkor in Malaya: The Clinical Evidence (Part I)," *Journal of Tropical Pediatrics*, June 1961, 3–15, and Dean, "Kwashiorkor in Malaya: The Clinical Evidence (Part II)," *Journal of Tropical Pediatrics*, September 1961, 39–48; "Child Research Unit-Future of the Unit, I," FD 12/274 PRO; and "Future of Dr. R.G. Whitehead," July 24, 1963, FD 12/274 PRO.

63. Whitehead to Lush, November 30, 1964, FD 12/274 PRO.

64. "Protein Calorie Malnutrition," in *Medical Care in Developing Countries: A Primer on the Medicine of Poverty and a Symposium from Makerere*, ed. Maurice King (Nairobi, Ken.: Oxford University Press, 1966), p. 14:14.

65. Ibid.

66. Ibid.

67. Ibid.

68. D. C. Robinson, "The Nutrition Rehabilitation Unit at Mulago Hospital, Kampala: Further Development and Evaluation 1967–69," in "Recent Approaches to Malnutrition in Uganda" Monograph No. 13 *Journal of Tropical Pediatrics and Environmental Child Health* 17, no. 1 (March 1971): 37.

69. Jelliffe and Dean, "Protein-Calorie Malnutrition, " 98; Dean and Jelliffe, "Treatment of Protein-Calorie Malnutrition," 429–39.

70. Jelliffe and Dean, "Protein-Calorie Malnutrition," 103.

71. Welbourn, "Child Welfare in Mengo District," 29.

72. Ainsworth, *Infancy in Uganda*, 23.

73. Minutes of Meeting Held at the Bishop's House, Namirembe, August 9, 1957, "To Consider the Acceptance and Distribution of Gifts from Church World Service" and Minutes of Welfare Committee, August 20, 1957, Uganda Christian University (UCU) Archives RG 1/BP/187/3.

74. Interview with Joyce Lukwago Nakidalu, Kisaaku, Uganda, July 21, 2012; and interview with Daisy Nakyejwe, Luteete Uganda, July 20, 2012.

75. Ainsworth, *Infancy in Uganda*, 62–63.

76. Ibid., 68.

77. H. F. Welbourn, "Bottle Feeding: A Problem of Modern Civilization," *Journal of Tropical Pediatrics*, March 1958, 163.

78. Welbourn, "Weaning among the Baganda," 17.

79. Josephine M. Namboze, "Weaning Practices in Buganda," *Tropical and Geographical Medicine* 19 (1967): 155.

80. Ainsworth, *Infancy in Uganda*, 303–7.

81. Ibid.

82. Ibid., 296–302.

83. Ibid., 300.

84. Ibid.

85. Ibid., 224.

86. Interview with Catherine Nansamba, Luteete, Uganda, July 20, 2012.

87. Welbourn, "Child Welfare in Mengo District." Clinics were held in Kampala and in "peri-urban" villages within twenty miles of the capital on a weekly, biweekly, or monthly schedule, with 7,719 children attending in 1954 alone.

88. Welbourn, "Bottle Feeding,"157.

89. Ainworth, *Infancy in Uganda*, 66.

90. Ibid., 158.

91. Musoke, "Analysis of Admissions," 310.

92. Namboze, "Weaning Practices in Buganda," 154–56.

93. Welbourn, "Weaning among the Baganda," 17.

94. Ainsworth, *Infancy in Uganda*, 285.

95. Nansamba, interview.

96. Musoke, "Analysis of Admissions," 314; "Appendix: Subjects that might Be Included in a Programme of Nutritional Research in Adults in Kampala," from Dean to Himsworth, January 18, 1963, FD 12/274 PRO; and D. B. Jelliffe, "The Need for Health Education," in *Health Education and the Mother & Child in East Africa*, Report of a Seminar Organised by the Departments of Paediatrics & Child Health, and of Preventive Medicine, with the co-operation of UNICEF, at Makerere Medical School, Kampala Uganda, November 12th-16th, 1961, Albert Cook Library (hereafter ACL).

97. Ainsworth, *Infancy in Uganda*, 69.

98. Jelliffe, "Need for Health Education," 1. Welbourn cited similar figures in Hebe F. Welbourn and Grace De Beer, "Trial of a Kit for Artificial Feeding in Tropical Village Homes," *Journal of Tropical Medicine and Hygiene* 67, no. 155 (1964): 155.

99. Michael C. Latham, "International Perspectives on Weaning Foods: The Economic and Other Implications of Bottle Feeding and the Use of Manufactured Weaning Foods," in *Breastfeeding and Food Policy in a Hungry World*, ed. Dana Raphael, (New York: Academic Press, 1979), 121.

100. Welbourn, "Weaning among the Baganda,"17; and Welbourn, "Bottle Feeding,"164.

101. Welbourn, "Bottle Feeding," 162 and 165.

102. Ibid., 161–62.

103. Ibid., 165.

104. Musoke, "Analysis of Admissions," 309–10.

105. H. C. Trowell and D. B. Jelliffe, eds., *Diseases of Children in the Subtropics and Tropics* (London: Edward Arnold, 1958), 118–19.

106. Derrick B. Jelliffe and E. F. Patrice Jelliffe, *Human Milk in the Modern World: Psychosocial, Nutritional and Economic Significance* (Oxford: Oxford University Press, 1979), 234.

107. Wrigley, "Changing Economic Structure," 64; Musoke, "Analysis of Admissions," 305; and Grace Bantebya Kyomuhendo and Marjorie Keniston McIntosh, *Women, Work and Domestic Virtue in Uganda, 1900–2003* (Athens: Ohio University Press, 2006), 13. This figure is notably for the entire Protectorate of Uganda and if a comparable figure were available for Buganda alone it is reasonable to assume that it would be higher. Some estimated that the population of Kampala was "likely to be over 100,000" (Southall and Gutkind, *Townsmen in the Making*, 8).

108. David E. Apter, *The Political Kingdom in Uganda: A Study in Bureaucratic Nationalism* (New Jersey: Princeton University Press, 1961), 40.

109. Wrigley, "Changing Economic Structure." The rising income of cash crop farmers in this period was still limited by the policy of fixing prices in order to cushion farmers from price fluctuations. In the early 1950s, revised policies and subsiding wartime inflation meant that even though farmers did not receive the full market value of their products, individual incomes did rise.

110. Richards, *Tribal Change*; A. B. Mukwaya, *Land Tenure in Buganda: Present Day Tendencies* (Kampala, Uga.: Eagle Press, 1953); Richards, Sturrock, and Fortt, eds., *Subsistence to Commercial Farming*; and West, *Land Policy*.

111. L. A. Fallers, assisted by F. K. Kamoga and S. B. K. Musoke, "Social Stratification in Traditional Buganda," in Fallers, *King's Men*, 105–6.

112. Whitehead, interview.

113. Jelliffe and Jelliffe, *Human Milk*, 239–40.

114. Donald S. McLaren, "The Great Protein Fiasco," *Lancet* 304 (July 1974): 93–96.

115. McLaren, "Great Protein Fiasco," 95.

116. Donald S. McLaren, "The Great Protein Fiasco Revisited," *Nutrition* 16 (2000): 464. For the original articulation of this argument, see McLaren, "Great Protein Fiasco," 95.

117. "Correspondence re Kwashiorkor - with DS McLaren, 1972–1975," 1980, PP/CDW, Box 32: H.2/1, WL.

118. McLaren, "Great Protein Fiasco," 95.

119. Jelliffe left Uganda in 1967 to direct the Caribbean Food and Nutrition Institute.

120. John Dobbing, ed., *Infant Feeding: Anatomy of a Controversy, 1973–1984* (New York: Springer-Verlag, 1988), 29–35.

121. D. B. Jelliffe, "Commerciogenic Malnutrition," *Food Technology* 25 (1971): 154.

122. Hugh Geach, "The Baby Food Tragedy," *New Internationalist*, no. 006 (August 1973), 8–12 and 23; and Mike Muller, *The Baby Killer: A War on Want Investigation into the Promotion and Sale of Powdered Baby Milks in the Third World* (London: War on Want, 1974).

123. Muller, *Baby Killer*, 1.

124. Muller, *Baby Killer*, 5 and 7.

125. Dobbing, *Infant Feeding*.

126. Nancy Rose Hunt, "'Le Bebe en Brusse': European Women, African Birth Spacing and Colonial Intervention in Breast Feeding in the Belgian Congo," *International Journal of African Historical Studies* 21, no. 3 (1988): 401–32.

Chapter 3: The Miracle of Kitobero

1. J. P. Stanfield, "Introduction," in "Recent Approaches to Malnutrition in Uganda," Monograph No. 13 *Journal of Tropical Pediatrics and Environmental Child Health* 17, no. 1 (March 1971): 3.

2. Note that the rise of bottle feeding discussed in the preceding chapter was not recognized as a factor inhibiting prevention as biomedical personnel failed to connect their distribution of skim milk and their skim milk–based curative formulas to the growing use of nursing bottles in Uganda.

3. J. P. Stanfield, "A Proposal that Dr. J.P. Stanfield Joins the Medical Research Council's Child Nutrition Research Unit on a Full Time Basis," November 21, 1968 (date of adjoining correspondence: Stanfield to Lush), FD 12/276, PRO.

4. Mahmoud Mohamed Hassan, "Health Education for Young Children in the Sudan," in "*Health Education,*" ACL, 21. Emphasis added.

5. Jelliffe and Dean, "Malnutrition in Early Childhood"; and Stanfield, "A Proposal," FD 12/276, PRO.

6. Church and Stanfield, interview.

7. Paget Stanfield, "A Project for the Evaluation of Nutrition Rehabilitation Methods as Developed and Practiced at the Nutrition Rehabilitation Unit (S.C.F.) Makerere Medical School, Kampala, Uganda," 1970, Personal Papers of Dr. Paget Stanfield. Emphasis in original.

8. F. J. Bennett, G. A. Saxton, J. S. W. Lutwama, and J. M. Namboze, "The Use of a Rural Community in the Curriculum of a Medical School in a Developing Region (East Africa)," 1965, p. 1–2, folder 23, box 2, series 492-A-Uganda, Record Group 1.2, Rockefeller Foundation Archives (hereafter RAC).

9. For more on Mengo Hospital see chapter 1.

10. Interview with Dr. Josephine Namboze, Kampala, Uganda, July 12, 2004.

11. Josephine M. Namboze, "Mobile Young Children's Clinics in Kasangati Health Centre Defined Area," *Journal of Tropical Pediatrics* 12 (Supplement 3) (1966): 1–2.

12. Dr. Loeb diary excerpt, September 16, 1963, folder 23, box 2, series 492, RG 1.2, RAC.

13. Namboze, interview.

14. Prior to his arrival in Uganda, Jelliffe worked in Sudan, Nigeria, India, the West Indies, and New Orleans. See for example: D. B. Jelliffe, *Infant Nutrition in the*

Subtropics and Tropics, World Health Organization Monograph Series No. 29 (Geneva: WHO, 1955); Jelliffe, "Social Culture and Nutrition: Cultural Blocks and Protein Malnutrition in Early Childhood in Rural West Bengal," *Pediatrics,* July 1957, 128–38; and Jelliffe, "Protein-Calorie Malnutrition in Tropical Preschool Children: A Review of Recent Knowledge," *Journal of Pediatrics* 54 (1959): 227–56. Jelliffe's approach to pediatrics in Uganda "set a pattern for other medical schools in Africa and in other parts of the world. It is fair to say that he personally educated and trained a cohort of pediatricians from almost every East African country, most of whom returned to their countries to become leaders and professors of pediatrics and child health, setting up departments in their own medical schools." He continued this work as the Chair of Public Health and Pediatrics and the Director of the International Public Health Program at the University of California, Los Angeles. "Derrick B. Jelliffe, Public Health: Los Angeles" in *University of California: In Memoriam, 1993,* ed. David Krogh (Berkeley: University of California, 1993). http://content.cdlib.org/view?docId=hb0h4n99rb&doc.view=frames&chunk.id=div00034&toc.depth=1&toc.id=&brand=calisphere.

15. D. B. Jelliffe, "Custom and Child Health in Buganda: I. Introduction," *Tropical and Geographical Medicine* 15 (1963): 121–23.

16. D. B. Jelliffe et al., "Field Survey of the Health of Bachiga Children in the Kayonza District of Kigezi, Uganda," *American Journal of Tropical Medicine and Hygiene* 10 (1961): 435–45; D. B. Jelliffe et al., "The Children of the Lugbara: A Study in the Techniques of Paediatric Field Survey in Tropical Africa," *Tropical and Geographical Medicine* 14 (1962): 33–50; D. B. Jelliffe et al., "The Health of Acholi Children," *Tropical and Geographical Medicine* 15 (1963): 411–12; and D. B. Jelliffe et al., "Ecology of Childhood Disease in the Karamojong of Uganda," *Archives of Environmental Health and Childhood Disease* 9 (1964): 25–36.

17. "Project 4: A Project for the Evaluation of Nutrition Rehabilitation Methods as Developed and Practiced at the Nutrition Rehabilitation Unit (S.C.P.), Pediatric Department, Makerere Medical School and at Luteete Rural Maternity and Child Health (Family) Clinic," adjoining correspondence Stanfield to Lush, November 21, 1968, FD 12/276, PRO; and Church and Stanfield, interview.

18. Namboze, interview.

19. Paget Stanfield, "The Save the Children Fund Nutrition Unit," December 7, 1965, Personal Papers of Dr. Paget Stanfield.

20. "Editorial: Nutritional Rehabilitation and Tropical Maternal and Child Health," *Journal of Tropical Pediatrics and African Child Health* 6, no. 2 (1960): 35. Professor Jelliffe was the founding and continuing editor of the *Journal of Tropical Pediatrics.*

21. Interview with Dr. John Kakitahi, Makerere Medical School, Mulago Hospital, Kampala, June 23, 2004.

22. Schneideman, Bennett, and Rutishauser, "Nutrition Rehabilitation Unit at Mulago," 29.

23. Early analysis for this project interpreted the education of mothers in a far more negative light in two respects. First, I viewed Mwanamugimu as an extension

of Christian missionary efforts to inculcate Victorian ideals of mothercraft, and as the kind of social reform and betterment campaign that has been widely written about in Africa. See, for example: Karen Tranenberg Hansen, ed., *African Encounters with Domesticity* (New Brunswick, NJ: Rutgers University Press, 1992); Deborah Gaitskell, "Housewives, Maids or Mothers: Some Contradictions of Domesticity for Christian Women in Johannesburg, 1903–1939," *Journal of African History* 24 (1983): 241–56; and more recently, Abosede A. George, *Making Modern Girls: A History of Girlhood, Labor, and Social Development in Colonial Lagos* (Athens: Ohio University Press, 2014). Further reflection led me to conclude that perhaps the program did not fit as easily within such a framework and may have instead consciously sought to distance itself from such paternalistic work. Secondly, the diminishing role of doctors and the increasing role of mothers that was so central to the demedicalization of malnutrition at Mwanamugimu involved a gendered reading of medicine as a masculine space and activity and the prevention of malnutrition as a maternal, feminine, less medical space and activity. This latter point warrants further research, which I ultimately decided could not be adequately pursued here.

24. "Project 4," FD 12/276, PRO.

25. "Project 4," FD 12/276, PRO, and Church and Stanfield, interview.

26. M. A. Church, "Some Ideas for Discussion Relating to 'Some Newer Community Approaches," August 1971, Personal Papers of Dr. Paget Stanfield; Stanfield, "Save the Children Fund,"; and Ian Schneideman, "'Mwanamugimu' Clinic," February 1967, Personal Papers of Dr. Mike Church.

27. Mike Church was the son of the medical missionary Joe Church and was born in Uganda. He was appointed the medical officer at the Nutrition Rehabilitation Unit after returning to Uganda following the completion of his medical training in London in 1966. Church, "Some Ideas for Discussion."

28. "Project 4," FD 12/276, PRO.

29. It is also worth noting that in 1967, 82 percent of scheduled appointments were kept and in 1965 the figure was 89 percent (Robinson, "Further Development and Evaluation 1967–69," 36).

30. Schneideman, "'Mwanamugimu' Clinic."

31. Church and Stanfield, interview. Several of the former wards have since been converted back into pediatric wards, one of which is now called the "Stanfield Children's Ward."

32. Schneideman, "'Mwanamugimu' Clinic."

33. They were prevented from doing so by the Ugandan Ministry of Health (Church and Stanfield, interview).

34. Schneideman, Bennett, and Rutishauser, "Nutrition Rehabilitation Unit at Mulago," 28.

35. Church and Stanfield, interview.

36. Stanfield, "Introduction," 3.

37. J. P. Stanfield, "Nutrition Education in the Context of Early Childhood Malnutrition in Low-Resource Communities," *Proceedings of the Nutrition Society* 35 (1976): 134.

38. A close analysis of the proverb, *mwanamugimu ava ku ngozi,* reveals a number of subtle meanings embedded within this particular axiom outlining the role of motherhood in Ganda society. The final word of the proverb, ngozi, refers specifically to the barkcloth or sling in which mothers carried infants and young children on their backs prior to weaning. Ngozi also signifies the amniotic membranes in which an unborn fetus develops. If the membranes don't break and were not broken during childbirth, an infant might be born still enclosed within the amniotic sac or ngozi. In Ganda cultural ideology at the turn of the century, when so many of our earliest sources were written, the role of mothers in child health and welfare was predominantly centered on the period of a child's life that preceded weaning and the naming ceremony. As part of the weaning process, Ganda mothers not only ceased breastfeeding but no longer slept in the same bed or carried their newly weaned children in a sling or ngozi. Children who had been weaned were also ideally sent to live with and be raised by relatives, and preferably the paternal aunt (*senga*) or grandmother. See Mair, *An African People,* 56; and Hebe F. Welbourn, "Custom and Child Health in Buganda: II. Methods of Child Rearing," *Tropical and Geographical Medicine* 15 (1963): 128. Weaning in Ganda society, therefore, signaled far more than just an end to breastfeeding; it marked the end of this important phase of Ganda motherhood. It is thus that ngozi is translated as a direct reference to motherhood, specifically to the period from conception through weaning in which the life or health (*-mugimu*, which translates as good, or beautiful, or healthy) of a developing fetus and young child (*omwana*) was seen as dependent upon close and constant contact with the mother. For more on cultural ideologies of motherhood as it relates to the Mwanamugimu program, see Jennifer Tappan, "'A Healthy Child Comes From A Healthy Mother': Mwanamugimu and Nutritional Science in Uganda, 1935–1973" (Ph.D. thesis, Columbia University, 2010). For more on the early history of motherhood in Uganda, see Rhiannon Stephens, *A History of African Motherhood: The Case of Uganda, 700–1900* (Cambridge: Cambridge University Press, 2013).

39. Church and Stanfield, interview.

40. Dean and Jelliffe, "Protein-Calorie Malnutrition," 431.

41. Church and Stanfield, interview.

42. Jelliffe, "I. Introduction," 121; and D. B. Jelliffe and F. J. Bennett, "Indigenous Medical Systems and Child Health," *Journal of Tropical Pediatrics* 57, no. 2 (August 1960): 248.

43. Church and Stanfield, interview.

44. Welboum, "Reminiscences of My Career," RHL, MSS.Afr.s.1872(152), 8.

45. Welbourn, "Child Welfare in Mengo District," 29–31.

46. As Mike Church pointed out, the use of sieves was, not unlike the promotion of bottle feeding, hazardous for the health of very young children who lived in areas without sufficient access to clean water supplies (Church and Stanfield, interview).

47. Welbourn, "Reminiscences of My Career," RHL, MSS.Afr.s.1872(152), 8; and Welboum, "Child Welfare in Mengo District," 31. The Luganda booklet was

entitled "Endiisa Enungi Omwana." The booklet "How to Feed Your Child" was published by Eagle Press in Nairobi in 1954 and many of the recipes were also included in her "Health in the Home" booklet that was later published by the East African Literature Bureau.

48. Luwombo refers to a banana leaf that has been prepared specifically to wrap and steam the ingredients of several Ganda dishes. These are dishes that are steamed like matooke, but because all of the ingredients are mixed to form a soup prior to the steaming process, greater care must be taken in selecting and preparing a young banana leaf that is not torn and will not allow the contents to seep out during the steaming process.

49. D. B. Jelliffe, C. Morton, and G. Nansubuga, "*Ettu* Pastes in Infant Feeding in Buganda," *Journal of Tropical Medicine and Hygiene* 65 (1962): 43.

50. M. A. Church, "The Background and Contents of Calendar Yamaka 1969," Personal Papers of Dr. Mike Church.

51. M. A. Church, "Songs for Influence in Nutrition Education," February 1970, Personal Papers of Dr. Mike Church.

52. Ekitobero is not actually a term found in Luganda dictionaries from the period and the complicated etymology is, therefore, part of the concept's rich idiom (Snoxall, *Luganda-English Dictionary*).

53. Interview with Patrick Njovu, New York, New York, January 9, 2007. The possibility that the preflx eki carried positive connotations is surmised from J. D. Chesswas, *The Essentials of Luganda* (1954; East African Literature Bureau Edition, Nairobi: Oxford University Press, 2002). Mike Church repeatedly stressed that ekitobero was thought of as an "excellent" food (Church and Stanfield, interview). See also M. A. Church, "Educational Methods, Cultural Characteristics and Nutrition Rehabilitation: Experience in Kampala Unit," in "Recent Approaches to Malnutrition in Uganda." Monograph No. 13 *Journal of Tropical Pediatrics and Environmental Child Health* 17, no. 1 (March 1971): 45; and Church, "Contents of Calendar Yamaka 1969."

54. Stanfleld to Lush, December 24, 1968, FD 12/276, PRO. Stanfield spells it *akabokisi*. Request to use "sauce pans which have got lids on to instead of Luwombo?" was recorded as part of a lesson on ettu paste preparation: "Kasangati Health Centre Community Health Record," June 28, 1963, folder 24, box 2, series 492-A-Uganda, RG 1.2, RAC.

55. Interview with Maria Zerenah Namusoke, Kisanku, Uganda, July 24, 2012.

56. Interview with Peragiya Nanziri, Bamunanika, Uganda, July 19, 2012.

57. Interview with Madina Namakula, Kisanku, Uganda, July 24, 2012.

58. Interview with Fatuma Nankabirwa, Bamunanika, Uganda, July 19, 2012.

59. Nansamba, interview.

60. Interview with Solom Nanyonga, Luteete, Uganda, July 19, 2012.

61. Stanfield, "Evaluation of Nutrition Rehabilitation Methods."

62. Church, "Some Ideas for Discussion."

63. Kakitahi, interview.

64. Schneideman, Bennett, and Rutishauser, "Nutrition Rehabilitation Unit at Mulago," 28. Arrangements were made, for example, to invite the "village chief" as a well as neighbors to a demonstration "given by the mother aided by the Unit's educator."

65. Church, "Contents of Calendar Yamaka 1969."

66. *A Play about Kitobero*, Personal Papers of Dr. Mike Church.

67. Church, "Songs for Influence"; and Church and Stanfield, interview. Kadongo kamu is figurative for "one guitar," and was a very popular form of ballad music beginning in the 1960s. Dan Mugula, who is credited as one of the "oldest and most prominent" kadongo kamu players and the *Jjajja w'abayimbi* or grandfather of Ugandan musicians, released his first recording in 1962. See Sylvia Nannyonga-Tamusuza, "Constructing the Popular: Challenges of Archiving Ugandan 'Popular' Music," *Current Writing: Text and Reception in Southern Africa* 18, no. 2 (2006): 33 52; and Elvis Basudde, "Uganda's Forgotten Music Heroes," *New Vision*, Tuesday, January 22, 2008.

68. Church, "Songs for Influence." Emphasis in original removed.

69. Ibid. The translation is provided by M. A. Church. Emphasis in original.

70. *A Play about Kitobero.*

71. Ibid.

72. *The Story of the Builder*, Personal Papers of Dr. Mike Church.

73. Ibid.

74. M. A. Church and J. P. Stanfield, "The Weight Chart: An Invaluable Aid in Nutrition Rehabilitation," Monograph No. 13 *Journal of Tropical Pediatrics and Environmental Child Health* 17, no. 1 (March 1971): 61.

75. Church and Stanfield, interview.

76. Church and Stanfield, "Weight Chart," 62.

77. Ibid., 62.

78. Ibid., 63.

79. Church, "Educational Methods," 52.

80. Ibid., 51.

81. M. A. Church, "Mwanamugimu," *Mother and Child*, May/June 1970, 14, personal papers of Dr. Mike Church.

82. Stanfield, "The Save the Children Fund"; Church, "Some Ideas for Discussion"; and Schneideman, "'Mwanamugimu' Clinic."

83. Aili Mari Tripp, *Women and Politics in Uganda* (Madison: University of Wisconsin Press, 2000), 28–29.

84. Ibid., 34–35.

85. "Minutes Medical Subconference," July 13, 1939, M/Y A7/10 CMS; "Medical Sub-Conference Minutes," M/Y A7/10 CMS; "Minutes L.C.M.T.S. Committee," July 11, 1939, M/Y A7/10 CMS; "Mengo Hospital: Minutes of the Medical Subconference," February 17, 1939, M/Y A7/10 CMS; and "Notes on Meeting of Medical Subconference," February 17, 1939, M/Y A7/10 CMS. For more on the midwifery training program see: Vaughan, *Curing Their Ills*; Vaughan, "Syphilis in

Colonial East Africa"; Summers, "Intimate Colonialism"; Tuck, "Venereal Disease, Sexuality and Society"; Musisi, "Politics of Perception,"; and Doyle, *Before HIV.*

86. Ndugwa, interview.

87. Church and Stanfield, interview; and J. P. Stanfield, "The Luteete Family Health Centre: Nutrition Rehabilitation in a Comprehensive Rural Development Strategy," *Journal of Tropical Pediatrics and Environmental Child Health* 17, no. 1 (March 1971): 68.

88. Interview with Kasifa Kyeyunne, Butto, Uganda, May 21, 2004; and interview with Florence Joyce and Wilson Ssalongo Kyaze, Luteete, Uganda, June 2, 2004.

89. "Application for Financial Support for a Pilot Study of the Nutrition Rehabilitation Unit," November 1964 folder Makerere College (Medical School), series 492, Record Group 2, RAC.

90. Whitehead, "Kwashiorkor in Uganda," 310.

91. Robinson, "Further Development and Evaluation 1967–69," 36.

92. E. R. Watts and D. G. R. Belshaw, "The Small Holding at Luteete Family Health Centre," *Journal of Tropical Pediatrics and Environmental Child Health* 17, no. 1 (March 1971): 83–87.

93. Whitehead, "Kwashiorkor in Uganda," 310–11.

94. Interview with Bumbakali Kyeyunne, Butto, Uganda, July 21, 2004.

95. Interview with Nabanja Federesi Kaloli, Luteete, Uganda, June 17, 2004; and B. Kyeyunne, interview.

96. Stanfield, "Luteete Family Health Centre," 77–78; and Church and Stanfield, interview.

97. Joyce and Kyaze, interview; Luteete, interview; and Kaloli, interview.

98. Namakula, interview.

99. Church and Stanfield, interview.

100. Joyce and Kyaze, interview.

101. Kaloli, interview; and K. Kyeyunne, interview.

102. Interview with Robinah Namulindwa Kayemba, Luteete, Uganda, July 23, 2004.

103. Interview with Kasifa Kyeyunne, Butto, Uganda, May 21, 2004; Interview with Florence Joyce and Wilson Ssalongo Kyaze, Luteete, Uganda, June 2, 2004.

104. Interview with Kasifa Kyeyunne, Butto, Uganda, May 21, 2004. Her eldest son, now an adult with his own family and children, was present during the interview and when discussing *kitobero* she pointed to him and said that she gave "this one *kitobero* for five years!"

105. Interview with Robinah Namulindwa Kayemba, Luteete, Uganda, July 23, 2004. In later interviews with residents in Luteete who were not members of the Tusitukirewamu club, I posed more general questions at the outset and only later in the interview made my specific interest in Mwanamugimu known. Robinah Kayemba began to discuss Mwanamugimu and kitobero in response to a general question about infant feeding.

106. Interview with Robinah Nanteza, Luteete, Uganda, July 2004.

107. Interview with Kasalina Nadamba, near Luteete Uganda, May 26, 2004; and interview with Daisy Nakyejwe, Luteete, Uganda, July 23, 2004.

108. Nanteza, interview.

109. Interview with Keziya Kazibwe, Luteete, Uganda, June 16, 2004.

110. Florence was not the wife of Luka Mukasa, as inaccurately indicated in Jennifer Tappan, "'A Healthy Child."

111. Namusoke, interview.

112. Nanziri, interview.

113. Nansamba, interview.

114. Interview with "Ssalongo" Stephen Mulindwa Maseruka, July 17, 2012.

115. Kayemba, interview.

116. Interview with Budesiaa Namusoke, Luteete, Uganda, July 24, 2004. Budesiaa's daughter was particularly striking, as she reported learning how to prepare kitobero from Florence Mukasa in the 1990s, long before giving birth to her twins, who were just beginning to eat solid foods at the time of our interview.

Chapter 4: In the Shadows of Structural Adjustment and HIV

1. Interview with Ephraim Musoke, Luteete, Uganda, May 13, 2004.

2. Julie Livingston, *Improvising Medicine: An African Oncology Ward in an Emerging Cancer Epidemic* (Durham, NC: Duke University Press, 2012).

3. James Ferguson, *Global Shadows: Africa in the Neoliberal World Order* (Durham, NC: Duke University Press, 2006), 40; and Alicia Decker, *In Idi Amin's Shadow: Women, Gender and Militarism in Uganda* (Athens: Ohio University Press, 2014).

4. V. F. Amann, D. G. R. Belshaw, and J. P. Stanfield, eds. *Nutrition and Food in an African Economy*, 2 vols. (Kampala: Makerere University, 1972); and "Recent Approaches to Malnutrition in Uganda" Monograph No. 13 *Journal of Tropical Pediatrics and Environmental Child Health* 17, no. 1 (March 1971).

5. Whitehead, interview.

6. R. G. Whitehead and J. P. Stanfield, "Proposals for the Child Nutrition Research Unit, Kampala, Uganda" FD 12/276, PRO; "Press Release - Opening Day," July 14, 1969, FD 12/281, PRO; Whitehead, interview; R. G. Whitehead, "Report on the Research Activities and Plans of R. G. Whitehead in Cambridge and Uganda," adjoining correspondence: Whitehead to Lush, February 7, 1966, FD 12/275, PRO; Whitehead to Gardner, March 17, 1971, PRO FD 12/276; and "Progress Report," FD 12/282, PRO.

7. Whitehead, "Kwashiorkor in Uganda," 310–11.

8. Speech given by the Deputy Minister of Health, Uganda, Mr. S. W. Uringi, July 17, 1969, FD 12/281, PRO.

9. Eria Muwazi was at the forefront of this effort in the 1940s and 1950s (see Iliffe, *East African Doctors*; and chapter 1).

10. Iliffe, *East African Doctors*, 125 and 137.

11. Ibid., 118–25.

12. Whitehead to Gray, September 24, 1971, FD 12/276, PRO.

13. Iliffe, *East African Doctors*, 122–42.

14. Extract of Baker's Visit to E. Africa, 1971: Tuesday, January 26, FD 12/276, PRO.

15. "Report of Dr. J.A.B. Gray's Visit to Uganda," October 1970, FD 12/276, PRO; Gray to Whitehead, June 18, 1971, FD 12/276, PRO; and Whitehead to Murray, April 17, 1971, FD 12/276, PRO.

16. Whitehead, interview; Whitehead to Murray, April 17, 1971, FD 12/276, PRO; "2nd Draft of Principles of MRC in Further Conversations with University, Ministry of Health and National Council of Scientific Research Regarding the Creation of an Institute of Nutrition," May 7, 1971, FD 12/276, PRO; Whitehead to Gray, October 5, 1971, FD 12/276, PRO; and "Makerere Affair: Our Verdict," *People*, no. 867 (October 4, 1971), FD 12/276, PRO.

17. "First Draft: A Proposal for the Establishment in Uganda of a National Institute of Human Nutrition," adjoining correspondence: Whitehead to Gray, July 24, 1971, FD 12/276, PRO.

18. Stanfield, "Introduction," 3.

19. Whitehead, "Kwashiorkor in Uganda," 306.

20. Decker, *In Idi Amin's Shadow*, 44–47; and interview with Louis Mugambe Muwazi, Makerere Medical School, Mulago Hospital, Kampala, Uganda, March 2004.

21. Unnamed man interviewed with Robinah Nanteza, Luteete, Uganda, July 2004.

22. Church and Stanfield, interview; Musoke, interview, May 13, 2004; Nakyejwe, interview, July 23, 2004; and F. Kyaze and W. Kyaze, interview.

23. Samwiri Rubaraza Karugire, *Roots of Instability in Uganda* (Kampala, Uga.: Fountain Publishers, 1996), 58–59; and Decker, *In Idi Amin's Shadow*, 47.

24. Karugire, *Roots of Instability*, 72–77.

25. Holger Bernt Hansen, "Uganda in the 1970s: A Decade of Paradoxes and Ambiguities," *Journal of Eastern African Studies* 7, no. 1 (February 2013): 87–94.

26. Ibid; and Whitehead, interview.

27. Whitehead, interview.

28. Henry Kayemba, *A State of Blood: The Inside Story of Idi Amin* (New York: Ace Books, 1977), 40; and Iliffe, *East African Doctors*, 145.

29. Iliffe, *East African Doctors*, 146.

30. Interview with Dr. Alexander Odonga, Kampala, Uganda, 2004.

31. Iliffe, *East African Doctors*, 149.

32. Interview with Dr. Chris Ndugwa, Makerere Medical School, Mulago, Kampala, Uganda, June 8, 2004.

33. Iliffe, *East African Doctors*, 144–48.

34. See chapter 3.

35. Kakitahi, interview; and "Report on a Visit by R.G. Whitehead to the Child Nutrition Unit, Kampala, Uganda" October 1979, Personal Papers of Dr. Paget Stanfield.

36. Kakitahi, interview.

37. Maseruka, interview.

38. Iliffe, *East African Doctors*, 147.

39. Decker, *In Idi Amin's Shadow*, 221–22.

40. A. B. K. Kasozi, *The Social Origins of Violence in Uganda, 1964-1985* (Montreal: McGill-Queen's University Press, 1994), 121.

41. Kakitahi, interview.

42. Maseruka, interview; Kakitahi, interview.

43. Decker, *In Idi Amin's Shadow*, 140–46.

44. Godfrey B. Asiimwe, "From Monopoly Marketing to Coffee *Magendo*: Responses to Policy Recklessness and Extraction in Uganda, 1971–79," *Journal of Eastern African Studies* 7, no. 1 (February 2013): 104–24. This was true even though the magendo economy was a crucial source of wealth for Amin and his inner circle.

45. Iliffe, *East African Doctors*, 148; and Susan Reynolds Whyte, "Medicines and Self-Help: The Privatization of Health Care in Eastern Uganda," in *Changing Uganda: The Dilemmas of Structural Adjustment & Revolutionary Change*, ed. Holger Bernt Hansen and Michael Twaddle (Athens: Ohio University Press, 1991), 130–48.

46. The impact of this return to a subsistence economy is unclear and was undoubtedly uneven. The historian Holger Bernt Hansen has argued that "It had the effect that in this period of Ugandan history we saw little malnutrition and people in rural areas managed to keep their livelihood at a reasonable level" (Hansen, "Uganda in the 1970s," 98). Anecdotal evidence suggests that the Amin years were a time of heightened levels of malnutrition. During a visit to Uganda in 1979, Stanfield found that "'Obwosi' is back with many children and mothers sad and miserable – [even] some adults have a kwashiorkor appearance." P. Stanfield, "Uganda," diary of Paget's visit to Uganda, October 1979, Personal Papers of Dr. Paget Stanfield.

47. Decker, *In Idi Amin's Shadow*, 239–59; and Karugire, *Roots of Instability*, 84.

48. "3-H Escalates the Fight against Hunger," *Rotarian*, January 1984, 52. http://books.google.com/books?id=ETYEAAAAMBAJ&pg=PA52&dq=uganda+intitle:The+intitle:Rotarian&hl=en&sa=X&ei=Td5rU8DHoaSyAStxYGgDA&ved=0CHEQ6AEwEQ#v=onepage&q=uganda%20intitle%3AThe%20intitle%3ARotarian&f=false

49. Kakitahi, interview.

50. Ibid.

51. Kasozi, *Social Origins of Violence*, 136–43; and Karugire, *Roots of Instability*, 87–88.

52. The Luwero Triangle refers to the region bounded by the roads from Kampala to Gulu and to Hoima and the Kafu River. Nelson Kasfir, "Guerrillas and Civilian Participation: The National Resistance Army in Uganda, 1981–86," *Journal of Modern African Studies* 43, no. 2 (June 16, 2005): 271–96.

53. Kasozi, *Social Origins of Violence*, 4; Karugire, *Roots of Instability*, 90; Alastair Johnston, "The Luwero Triangle: Emergency Operations in Luwero, Mubende and Mpigi Districts," in *Crisis in Uganda: The Breakdown of Health Services*, ed. Cole P. Dodge and Paul D. Wiebe (New York: Pergamon Press, 1985), 101.

54. Johnston, "Luwero Triangle," 97–106; and Kasozi, *Social Origins of Violence*, 185.

55. Maseruka, interview.

56. Johnston, "Luwero Triangle," 105.

57. Kasozi, *Social Origins of Violence*, 174–75.

58. Ibid., 196–97.

59. Joshua B. Mugyenyi, "IMF Conditionality and Structural Adjustment Under the National Resistance Movement," in Hansen and Twaddle, *Changing Uganda*, 63.

60. K. Sarwar Lateef, "Structural Adjustment in Uganda: The Initial Experience," in Hansen and Twaddle, *Changing Uganda*, 37; Julius Kiiza, Godfrey Asiimwe, and David Kibikyo, "Understanding Economic and Institutional Reforms in Uganda," in *Understanding Economic Reforms in Africa: A Tale of Seven Nations*, ed. Joseph Mensah (New York: Palgrave Macmillan, 2007), 76.

61. Lateef, "Structural Adjustment in Uganda," 37; Mugyenyi, "IMF Conditionality and Structural Adjustment," 69.

62. Mugyenyi, "IMF Conditionality and Structural Adjustment," 70–71.

63. $301.2 million out of a total budget of $362.1 million in 1991–92, and a requested $333 million out of a total budget of $409.6 million for 1992–93, as cited in David Himbara and Dawood Sultan, "Reconstructing the Ugandan State and Economy: The Challenge of an International Bantustan," *Review of African Political Economy* 22, no. 63 (March 1, 1995): 91.

64. In Uganda, while national figures suggested progress in the fight against poverty, regional data indicate that over half of people living in rural areas (where the large majority of Ugandans live) continued in the early years of the new millennium to live below the poverty line, and that the actual number of people living in poverty only fell from an estimated 6.3 million in 1992 to 6 million in 2000 (Kiiza, Asiimwe, and Kibikyo, "Institutional Reforms in Uganda," 83).

65. James Pfeiffer and Rachel Chapman, "Anthropological Perspectives on Structural Adjustment and Public Health," *Annual Review of Anthropology* 39 (January 1, 2010): 149–65; Rachel Hammonds and Gorik Ooms, "World Bank Policies and the Obligation of Its Members to Respect, Protect and Fulfill the Right to Health," *Health and Human Rights* 8, no. 1 (January 1, 2004): 26–60; and Kiiza, Asiimwe, and Kibikyo, "Institutional Reforms in Uganda," 57–94.

66. Hammonds and Ooms, "World Bank Policies," 36; and Iliffe, *East African Doctors*, 146.

67. Dodge and Wiebe, *Crisis in Uganda*, xi.

68. Sam Agatre Okuonzi and Joanna Macrae, "Whose Policy Is It Anyway? International and National Influences on Health Policy Development in Uganda," *Health Policy and Planning* 10, no. 2 (1995): 125–26; Whyte et al., "Therapeutic Clientship"; James Pfeiffer, "International NGOs and Primary Health Care in Mozambique: The Need for a New Model of Collaboration," *Social Science & Medicine* 56, no. 4 (February 2003): 725–38; and Pfeiffer, "The Struggle for a Public Sector: PEPFAR in Mozambique," in Biehl and Petryna, *When People Come First*, 166–81.

69. Crane, *Scrambling for Africa*.

70. Iliffe, *East African Doctors*, 149–55; and Okuonzi and Macrae, "Whose Policy Is It Anyway?" 126–27.

71. The imposition of user fees was a policy, known as the Bamako Initiative, that was pursued across the continent in this time period with much the same results. See, for instance, Packard, *History of Global Health*, 263–64.

72. Okuonzi and Macrae, "Whose Policy Is It Anyway?" 127–29.

73. Whyte, "Medicines and Self-Help," 130–47. The salary figures are from Iliffe, *East African Doctors*.

74. Iliffe, *East African Doctors*, 165.

75. Whyte, "Medicines and Self-Help"; David Kyaddondo and Susan Reynolds Whyte, "Working in a Decentralized System: A Threat to Health Workers' Respect and Survival in Uganda," *International Journal of Health Planning and Management* 18, no. 4 (October 1, 2003): 329–42.

76. Harriet Birungi, "Injections and Self-Help: Risk and Trust in Ugandan Health Care," *Social Science & Medicine* 47, no. 10 (November 1998): 1455–62.

77. Whyte et al., "Therapeutic Clientship," 143; and Kiiza, Asiimwe, and Kibikyo, "Institutional Reforms in Uganda," 76.

78. Iliffe, *East African Doctors*, 158.

79. Ibid., 152–55.

80. Kiiza, Asiimwe, and Kibikyo, "Institutional Reforms in Uganda," 71–73; Sicherman, *Becoming an African University*, 261–63; and Iliffe, *East African Doctors*, 166–68.

81. Namboze, interview.

82. Kakitahi, interview.

83. Iliffe, *African Aids Epidemic*, 22–26.

84. Ibid., 71; Iliffe, *East African Doctors*, 220–34.

85. Iliffe, *African Aids Epidemic*, 99–101.

86. Crane, *Scrambling for Africa*; Iliffe, *African Aids Epidemic*, 149–50.

87. Crane, *Scrambling for Africa*, 120.

88. Ibid., 122.

89. Diana Nabiruma, "Mulago's Mwanamugimu Gets a Facelift," *Observer*, June 8, 2001, http://allafrica.com/stories/201106090983.html.

90. Nabiruma, "Mwanamugimu Gets a Facelift." For RESTORE International: http://restoreinternational.org/. For Mildmay: http://www.mildmay.org/.

91. Ugandanmedic, June 8, 2011, comment on Nabiruma, "Mwanamugimu Gets a Facelift."

92. Louise C. Ivers et al., "HIV/AIDS, Undernutrition, and Food Insecurity," *Clinical Infectious Diseases* 49, no. 7 (October 1, 2009): 1096–98.

93. Ibid., 1096.

94. Ibid.

95. Michael H. N. Golden and André Briend, "Treatment of Malnutrition in Refugee Camps," *Lancet* 342, no. 8867 (August 7, 1993): 360.

96. Steve Collins et al., "Management of Severe Acute Malnutrition in Children," *Lancet* 368, no. 9551 (December 8, 2006): 1994.

97. M. Namusoke, interview.

98. Nanziri, interview.

99. Namakula, interview.

100. Interview with Dorobina Kyambadde, Luteete, Uganda, July 20, 2012.

Epilogue

1. "United Nations Millennium Development Goals," United Nations, accessed August 21, 2014, http://www.un.org/millenniumgoals/.

2. André Briend, "Treating Malnutrition: New Issues and Challenges," in Crombé and Jézéquel, *A Not-So Natural Disaster*, 191.

3. "Nutriset's Timeline," Nutriset: Nutritional Autonomy for All, accessed August 21, 2014, http://www.nutriset.fr/en/about-nutriset/nutriset-timeline.html; and Rhichard Longhurst, "Famines, Food, and Nutrition: Issues and Opportunities for Policy and Research," *Food and Nutrition Bulletin* 9, no. 1 (1987), http://archive.unu.edu/unupress/food/8F091e/8F091E05.htm#Famines, food, and nutrition: issues and opportunities for policy.

4. For more on the famine and NGO work in the Sahel, see Gregory Mann, *From Empires to NGOs in the West African Sahel: The Road to Nongovernmentality*, African Studies Series 129 (New York: Cambridge University Press, 2015).

5. "Nutriset's Timeline," Nutriset.

6. See chapter 3; Dean, "Treatment of Kwashiorkor," 792; Dean and Weinbren, "Fat Absorption"; and Dean and Skinner, "On the Treatment of Kwashiorkor," 215–16.

7. World Health Organization, *Management of Severe Malnutrition: A Manual for Physicians and Other Senior Health Workers* (Geneva: WHO, 1999), http://www.who.int/nutrition/publications/severemalnutrition/9241545119/en/.

8. André Briend et al., "Ready-to-Use Therapeutic Food for Treatment of Marasmus," *Lancet* 353, no. 9166 (May 22, 1999): 1767–68.

9. "Nutriset's Timeline," Nutriset.

10. André Briend, "Highly Nutrient-Dense Spreads: A New Approach to Delivering Multiple Micronutrients to High-Risk Groups," *British Journal of Nutrition* 85, no. S2 (May 2001): S175; and Andrew Rice, "The Peanut Solution: Could a Peanut Paste Called Plumpy'nut End Malnutrition?" *New York Times*, September 2, 2010, sec. Magazine, http://www.nytimes.com/2010/09/05/magazine/05Plumpy-t.html.

11. Steve Collins and Kate Sadler, "Outpatient Care for Severely Malnourished Children in Emergency Relief Programmes: A Retrospective Cohort Study," *Lancet* 360, no. 9348 (December 7, 2002): 1824–30.

12. Briend, "New Issues and Challenges"; Isabelle Defourny, "Operational Innovation in Practice: MSF's Programme against Malnutrition in Maradi," in Crombé and Jézéquel, *A Not-So Natural Disaster*, 169–87.

13. Collins et al., "Severe Acute Malnutrition in Children," 1994.

14. Isabelle Defourny et al., "Management of Moderate Acute Malnutrition with RUTF in Niger," *Field Exchange* 31 (September 2007): 2; and Isabelle

Defourny et al., "A Large-Scale Distribution of Milk-Based Fortified Spreads: Evidence for a New Approach in Regions with High Burden of Acute Malnutrition," *PLoS ONE* 4, no. 5 (May 6, 2009): 1–7. For other studies of RUTF efficacy, see Michael A. Ciliberto et al., "Comparison of Home-Based Therapy with Ready-to-Use Therapeutic Food with Standard Therapy in the Treatment of Malnourished Malawian Children: A Controlled, Clinical Effectiveness Trial," *American Journal of Clinical Nutrition* 81, no. 4 (April 1, 2005): 864–70.

15. Defourny et al., "Milk-Based Fortified Spreads."

16. The term "hunger-wonder product" appears to have cropped up in the media coverage of the legal battle over Nutriset's patent; see, for example: Hugh Schofield, "Legal Fight over Hunger Wonder-Product," *BBC News*, April 8, 2010, sec. Europe, http://news.bbc.co.uk/2/hi/europe/8610427.stm; and "Our Cause," This Bar Saves Lives, accessed August 22, 2014, http://www.thisbarsaveslives.com/pages/our-cause. The conflation of malnutrition with hunger is problematic, particularly in association with a therapeutic product designed to combat severe acute malnutrition; see, for example: Jeffrey Sachs, Jessica Fanzo, and Sonia Sachs, "Saying 'Nuts' to Hunger," *Huffington Post* (blog), September 6, 2010, http://www.huffingtonpost.com/jeffrey-sachs/saying-nuts-to-hunger_b_706798.html.

17. Anderson Cooper, "'Miracle' Food Saves Lives," *60 Minutes* video, posted June 22, 2008, http://www.cbsnews.com/videos/miracle-food-saves-lives/.

18. Edesia, "Our Founder," accessed August 22, 2014, http://www.edesiaglobal.org/about-us/; and Rice, "Peanut Solution."

19. Rice, "Peanut Solution."

20. MANA, accessed August 22, 2014, http://mananutrition.org/.

21. Rice, "Peanut Solution."

22. *Community-Based Management of Severe Acute Malnutrition* (World Health Organization/World Food Programme/United Nations System Standing Committee on Nutrition/The United Nations Children's Fund, 2007), http://www.who.int/nutrition/topics/Statement_community_based_man_sev_acute_mal_eng.pdf; Isabelle Komrska, "Increasing Access to Ready-to-Use Therapeutic Foods (RUTF)," *Field Exchange* 42 (January 2012): 46; Martin Enserink, "The Peanut Butter Debate," *Science*, n.s., 322, no. 5898 (October 3, 2008): 36–38; and Project Peanut Butter, accessed August 22, 2014, http://www.projectpeanutbutter.org/.

23. Sheila Isanaka et al., "Effect of Preventive Supplementation with Ready-to-Use Therapeutic Food on the Nutritional Status, Mortality, and Morbidity of Children Aged 6 to 60 Months in Niger: A Cluster Randomized Trial," *Journal of American Medical Association* 301, no. 3 (January 21, 2009): 277–85; Lynnette M. Neufeld, "Ready-to-Use Therapeutic Food for the Prevention of Wasting in Children," *Journal of American Medical Association* 301, no. 3 (January 21, 2009): 327–28; and Enserink, "Peanut Butter Debate."

24. Ann Ashworth, "Efficacy and Effectiveness of Community-Based Treatment of Severe Malnutrition," *Food and Nutrition Bulletin* 27, no. S3 (2006): S41. For per-capita health expenditures: World Bank Group, "Health Expenditure Per Capita (Current US$)," World Health Organization Global Health Expenditure

database, accessed August 23, 2014, http://data.worldbank.org/indicator/SH.XPD.PCAP/countries?display=default .

25. Defourny, "Operational Innovation in Practice: MSF's Programme against Malnutrition in Maradi," 184–185.

26. Valerie Gatchell, Vivienne Forsythe, and Paul-Rees Thomas, "The Sustainability of Community-Based Therapeutic Care (CTC) in Non-acute Emergency Contexts," in SCN Nutrition Policy Paper No. S21, ed. Claudine Prudhon et al., *Food and Nutrition Bulletin* 27, no. 3 (September 2006): 6, http://www.who.int/nutrition/publications/severemalnutrition/FNB_0379_5721/en/ .

27. Defourny, "MSF's Programme against Malnutrition," 184; and Enserink, "Peanut Butter Debate," 36.

28. Cooper, "'Miracle' Food Saves Lives."

29. Rice, "Peanut Solution."

30. Ibid.

31. Stanfield, "A Proposal," FD 12/276, PRO.

GLOSSARY

akaboxi – small lidded pot
Baganda – people of the Buganda Kingdom (plural of Muganda)
bakopi – peasants
basawo – local healers (plural of musawo)
bwamaka bulungi – "beautiful home"
ebipimo – "measurements"
eccupa – bottle
ekigalanga – condition characterized by fever, diarrhea, abdominal pain, appetite loss, and cold feet; spiritual
emmere – "food"; usually a reference to the staple
empewo – wind, cold; condition related to swelling; spiritual
empindi – cowpeas, small bean
endwadde ez'ekiganda – Ganda diseases
endwadde ez'ekizungu – European/foreign diseases
enva – sauce, relish
ettu – packet/package
fundi – expert
jjajja – grandfather (*Jjajja w'abayimbi* or Grandfather of Ugandan singers/musicians)
kabaka – king
kadongo kamu – "one guitar" referring to a popular ballad type of music in Buganda
kibanja – garden plot (plural: *bibanja)*
kibuga – capital
kitobeko – "a mixture or variegated pattern"
kitobero – popular mixture of different kinds of foods
kwosera – to wean too soon
Luganda – language of the Baganda
lukiiko – gathering of chiefs, legislative council of Buganda
lumonde – sweet potato
lusuku – plaintain or banana garden
luwombo – banana leaves used in food preparation
magendo – smuggling
magezi – wisdom

mailo – land allotted as private property in early colonial period
matooke – plaintains, green cooking bananas (singular: *ettooke*)
miruka – district, parish
-mugimu – suffix meaning good, or beautiful, or healthy
nakati – leafy, green vegetable
ngozi – barkcloth sling used to carry young children
nkejji – small dried cichlid fish
obulwadde – illness, sickness, or disease
obusulo – childhood illness associated with swelling
obwosi – childhood illness associated with early weaning
olbuwadde bw'eccupa – "bottle disease"
olumbe – incurable disease, fatal illness
olutabu – soft and bland food
omusana – childhood illness associated with sun exposure
omwana – young child
senga – paternal aunt
tabula – stir up, mix
Tusitukirewamu – "let us all rise and help each other"

BIBLIOGRAPHY

Archival Sources

Albert Cook Library, Makerere Medical School, Mulago, Uganda
Child Health and Development Centre Library (CHDC), Mulago, Uganda
Church Missionary Society Archive (CMS), Cadbury Research Library, University of Birmingham
London School of Economics, London, UK
Makerere University Library, Kampala, Uganda
Ministry of Health, Kampala, Uganda
Public Record Office (PRO), The National Archives, London, UK
Rhodes House Library (RHL), Oxford University
Uganda Christian University (UCU) Archives, Mukono, Uganda
Uganda National Archives (UNA), Entebbe, Uganda
The Rockefeller Archive Center, Sleepy Hollow, NY
Wellcome Institute for the History of Medicine, London, UK

Personal Papers

Personal Papers of Dr. Mike Church
Personal Papers of Dr. Paget Stanfield

Unpublished Papers and Theses

Cooper, Barbara M. "Discourses on Infant Malnutrition and 'Bad Mothering' in a West African Setting." Paper presented at Health, Disease, and Environment in Africa: Histories of Current Challenges workshop, Princeton University, October 11, 2008.

Kodesh, Neil. "Beyond the Royal Gaze: Clanship and Collective Well-Being in Buganda." PhD thesis, Northwestern University, 2004.

Stephens, Rhiannon. "A History of Motherhood, Food Procurement and Politics in East-Central Uganda to the Nineteenth Century." PhD thesis, Northwestern University, 2007.

Summers, Carol. "Young Africa and Radical Visions: Revisiting the Bataka in Buganda, 1944–54." (unpublished paper, March 2004).

Tappan, Jennifer. "'A Healthy Child Comes from a Healthy Mother': Mwanamugimu & Nutritional Science in Uganda, 1935–1973." PhD thesis, Columbia University, 2010.

Tilley, Helen. "Africa as a 'Living Laboratory': The African Research Survey and the British Colonial Empire: Consolidating Environmental, Medical, and Anthropological Debates, 1920–1940." DPhil, Oxford University, 2001.

Tuck, Michael. "Syphilis, Sexuality, and Social Control: A History of Venereal Disease in Colonial Uganda." PhD thesis, Northwestern University, 1997.

Zeller, Diane. "The Establishment of Western Medicine in Buganda." PhD thesis, Columbia University, 1971.

Published Papers and Reports

"3-H Escalates the Fight against Hunger." *Rotarian*, January 1984, 52. http://books.google.com/books?id=ETYEAAAAMBAJ&pg=PA52&dq=uganda+intitle:The+intitle:Rotarian&hl=en&sa=X&ei=Td5rU8DHoaSyAStxYGgDA&ved=0CHEQ6AEwEQ#v=onepage&q=uganda%20intitle%3AThe%20intitle%3ARotarian&f=false.

Amann, V. F. D., G. R. Belshaw, and J. P. Stanfield, eds. *Nutrition and Food in an African Economy*. 2 vols. Kampala, Uga.: Makerere University, 1972.

Autret, M., and M. Behar. *Sindrome Policarencial Infantil (Kwashiorkor) and Its Prevention in Central America*. FAO Nutritional Studies 13. Rome: FAO, 1954.

Brock, J. F., and M. Autret. *Kwashiorkor in Africa*. World Health Organization Monograph Series 8. Geneva: WHO, 1952.

Colonial Office. *Malnutrition in African Mothers, Infants, and Young Children: Report of the Second Inter-African Conference on Nutrition, Fajara, Gambia, 19–27 November, 1952*. London: H. M. Stationery Office, 1954.

Committee on Protein Malnutrition, Food and Nutrition Board, National Research Council, and National Institutes of Health, Nutrition Study Section. *Progress in Meeting Protein Needs of Infants and Preschool Children: Proceedings of an International Conference*. Washington, DC: National Academy of Sciences, 1961.

Dean, R. F. A. *Plant Proteins in Child Feeding*. Medical Research Council Special Report Series 279. London: H. M. Stationery Office, 1953.

Department of Experimental Medicine, Cambridge, and associated workers, eds. *Studies of Undernutrition, Wuppertal 1946–9*. Medical Research Council Special Report Series 275. London: H. M. Stationery Office, 1951.

Economic Advisory Council, Committee on Nutrition in the Colonial Empire. *First Report—Part I. Nutrition in the Colonial Empire*. Cmd. 6050. London: H. M. Stationery Office, 1939.

———. *First Report—Part II. Summary of Information Regarding Nutrition in the Colonial Empire*. Cmd. 6051. London: H. M. Stationery Office, 1939.

FANTA-2. *The Analysis of the Nutrition Situation in Uganda*. Food and Nutrition Technical Assistance II Project (FANTA-2). Washington, DC: Academy for Educational Development, 2010. http://www.health.go.ug/hmis/public/nutrition/Uganda_Nutrition_Situation_Analysis.pdf.

Food and Agriculture Organization and World Health Organization. *Report of the FAO/WHO Seminar on Problems of Food and Nutrition in Africa South of the Sahara*. FAO Nutrition Meeting Report Series No. 25. Rome: N.p., 1961.

Jelliffe, D. B. *Infant Nutrition in the Subtropics and Tropics*. World Health Organization Monograph Series No. 29. Geneva: WHO, 1955.

"Joint FAO/WHO Expert Committee on Nutrition: Report on the First Session." World Health Organization Technical Report Series No. 16. Geneva: WHO, 1950.

"Joint FAO/WHO Expert Committee on Nutrition: Report on the Second Session." World Health Organization Technical Report Series No. 44. Geneva: WHO, 1951.

"Joint FAO/WHO Expert Committee on Nutrition: Report on the Third Session." World Health Organization Technical Report Series No. 72. Geneva: WHO, 1953.

"Joint FAO/WHO Expert Committee on Nutrition: Fifth Report." World Health Organization Technical Report Series No. 149. Geneva: WHO, 1958.

"Joint FAO/WHO Expert Committee on Nutrition: Sixth Report." World Health Organization Technical Report Series No. 245. Geneva: WHO, 1962.

McCance, R. A. "The History, Significance and Aetiology of Hunger Oedema." In Members of the Department of Experimental Medicine, *Wuppertal 1946–9*, 21–82.

Protein Advisory Group. *Lives in Peril: Protein and the Child.* World Food Problems No. 12. Rome: N.p., 1970.

Uganda Government, World Health Organization, and Food and Agriculture Organization. *Nutrition Seminar for English-Speaking Countries and Territories in Africa South of the Sahara.* FAO Report no. 960. Rome: FAO and WHO, 1958.

Waterlow, J. C., ed. *Protein Malnutrition, Proceedings of a Conference in Jamaica, Sponsored Jointly by the FAO, WHO and the Josiah Macy Jr. Foundation, New York.* Cambridge: Cambridge University Press, 1953.

Waterlow, J. C., H. C. Trowell, and J. N. P. Davies. "Liver Biopsy Demonstrations." In Colonial Office, *Malnutrition in African Mothers*, 100.

Waterlow, J. C., and A. Vergara. *Protein Malnutrition in Brazil.* FAO Nutritional Studies 14. Rome: FAO, 1956.

Widdowson, E. M. *A Study of Individual Children's Diets.* Medical Research Council Special Report Series No. 257. London: H. M. Stationery Office, 1947.

World Health Organization. *Essential Nutrition Actions: Improving Maternal, Newborn, Infant and Young Child Health and Nutrition.* Geneva: WHO, 2013.

———. *Management of Severe Malnutrition: A Manual for Physicians and other Senior Health Workers.* Geneva: WHO, 1999. http://www.who.int/nutrition/publications/severemalnutrition/9241545119/en/.

World Health Organization, the World Food Programme, the United Nations System Standing Committee on Nutrition, and the United Nations Children's Fund. "Community-Based Management of Severe Acute Malnutrition." 2007. http://www.who.int/nutrition/topics/Statement_community_based_man_sev_acute_mal_eng.pdf.

Published Scientific Articles

"Editorial: Nutritional Rehabilitation and Tropical Maternal and Child Health." *Journal of Tropical Pediatrics and African Child Health* 6, no. 2 (1960): 35.

Allison, James B. "Calories and Protein Nutrition." *Annals of the New York Academy of Sciences* 69 (1958): 1009–66.

Arya, O. P., and F. J. Bennett. "Health Education for University Students in Tropical Africa." *International Journal of Health Education* 12, no. 4 (1969): 164–69.

Ashworth, Ann. "Efficacy and Effectiveness of Community-Based Treatment of Severe Malnutrition." *Food and Nutrition Bulletin* 27, no. S3 (2006): S24–S48.

Béhar, Moisés, Fernando Viteri, Ricardo Bressani, Guillermo Arroyave, Robert L. Squibb, and Nevin S. Scrimshaw. "Principles of Treatment and Prevention of Severe Protein Malnutrition in Children (Kwashiorkor)." *Annals of the New York Academy of Sciences* 69 (1958): 954–68.

Bengoa, Jose M. "Nutrition Rehabilitation Centres." *Journal of Tropical Pediatrics* 13, no. 4 (1967): 169–76.

Bennett, F. J. "Custom and Child Health in Buganda: V. Concepts of Disease." *Tropical and Geographical Medicine* 15 (1963): 148–57.

———. "Health Education." In King, *Care in Developing Countries*, 6:1–6:13.

Bennett, F. J., and J. S. W. Lutwama. "Buganda Women: What They Want to Know." *International Journal of Health Education* 5, no. 2 (1962): 74–83.

Bennett, F. J., J. S. W. Lutwama, and Josephine Namboze. "Uganda: What Children Want to Know about Health." *International Journal of Health Education* 7, no. 2 (1964): 82–90.

Bennett, F. J., D. Serwadda, and D. B. Jelliffe. "Kiganda Concepts of Diarrhoeal Disease." *East African Medical Journal* 41, no. 5 (May 1964): 211–18.

Bhutta, Zulfiqar A., Jai K. Das, Arjumand Rizvi, Michelle F. Gaffey, Neff Walker, Susan Horton, Patrick Webb, Anna Lartey, and Robert E Black. "Evidence-Based Interventions for Improvement of Maternal and Child Nutrition: What Can Be Done and at What Cost?" *Lancet* 382 (August 3, 2013): 40–65.

Black, Robert E., Lindsay H. Allen, Zulfiqar A. Bhutta, Laura E. Caulfield, Mercedes de Onis, Majid Ezzati, Colin Mathers, and Juan Rivera. "Maternal and Child Undernutrition: Global and Regional Exposures and Health Consequences." *Lancet* 371 (January 19, 2008): 243–60.

Black, Robert E., Cesar G. Victora, Susan P. Walker, Zulfiqar A. Bhutta, Parul Christian, Mercedes de Onis, Majid Ezzati et al. "Maternal and Child Undernutrition and Overweight in Low-Income and Middle-Income Countries." *Lancet* 382 (August 3, 2013): 15–39.

Briend, André. "Highly Nutrient-Dense Spreads: A New Approach to Delivering Multiple Micronutrients to High-Risk Groups." *British Journal of Nutrition* 85, no. S2 (May 2001): S175–S179.

———. "Treating Malnutrition: New Issues and Challenges." In Crombé and Jézéquel, *A Not-So Natural Disaster*, 189–203.

Briend, André, Radandi Lacsala, Claudine Prudhon, Béatrice Mounier, Yvonne Grellety, and Michael H. N. Golden. "Ready-to-Use Therapeutic Food for Treatment of Marasmus." *Lancet* 353, no. 9166 (May 22, 1999): 1767–68.

Briend, André, Claudine Prudhon, Zita Weise Prinzo, Bernadette M. E. G. Daelmans, and John B. Mason. "Putting the Management of Severe Malnutrition Back on the International Health Agenda." *Food and Nutrition Bulletin*, 27, no. S3 (2006): S3–S6.

Briend, André, and James A Berkley. "Long Term Health Status of Children Recovering from Severe Acute Malnutrition." *Lancet Global Health* 4, no. 9 (2016): e590–e591. doi:10.1016/S2214-109X(16)30152-8.

Burgess, H. J. L. "The Medical Research Council's Rural Child Welfare Clinic." *East African Medical Journal*, May 1960, 391–98.

Church, M. A. "Educational Methods, Cultural Characteristics and Nutrition Rehabilitation: Experience in Kampala Unit," in Stanfield "Malnutrition in Uganda," 43–50.

Church, M. A., and J. P. Stanfield. "The Weight Chart: An Invaluable Aid in Nutrition Rehabilitation," in Stanfield, "Malnutrition in Uganda," 61–66.

Ciliberto, Michael A., Heidi Sandige, MacDonald J. Ndekha, Per Ashorn, André Briend, Heather M. Ciliberto, and Mark J. Manary. "Comparison of Home-Based Therapy with Ready-to-Use Therapeutic Food with Standard Therapy in the Treatment of Malnourished Malawian Children: A Controlled, Clinical Effectiveness Trial." *American Journal of Clinical Nutrition* 81, no. 4 (April 1, 2005): 864–70.

Clegg, K. Mary. "The Availability of Lysine in Groundnut Biscuits Used in the Treatment of Kwashiorkor." *British Journal of Nutrition* 14 (1960): 325–29.

Clegg, K. Mary, and R. F. A. Dean. "Balance Studies on Peanut Biscuit in the Treatment of Kwashiorkor." *American Journal of Clinical Nutrition* 8 (1960): 885–89.

Cole, T. J., and J. M. Parkin. "Infection and its Effect on the Growth of Young Children: A Comparison of The Gambia and Uganda." *Transactions of the Royal Society of Tropical Medicine and Hygiene* 71, no. 3 (1977): 196–98.

Collins, Steve, Nicky Dent, Paul Binns, Paluku Bahwere, Kate Sadler, and Alastair Hallam. "Management of Severe Acute Malnutrition in Children." *Lancet* 368, no. 9551 (December 8, 2006): 1992–2000.

Collins, Steve, and Kate Sadler. "Outpatient Care for Severely Malnourished Children in Emergency Relief Programmes: A Retrospective Cohort Study." *Lancet* 360, no. 9348 (December 7, 2002): 1824–30.

Cook, Albert R. "A Social Purity Campaign in Uganda." *Mercy and Truth*, 1921, 16–19.

———. "A Social Purity Campaign in Uganda." *Mission Hospital* 26 (1922): 32–35, 83–85, 110–12.

Davies, J. N. P. "The Essential Pathology of Kwashiorkor." *Lancet*, 1948, 317–20.

———. "The History of Syphilis in Uganda." *Bulletin of the World Health Organization* 15, no. 6 (1956): 1041–55.

Dean, R. F. A. "Biochemical Changes Caused by Protein Deficiency in Young Children." *Clinica Chimica Acta* 5 (1960): 186–91.

———. "Kwashiorkor in Malaya: The Clinical Evidence (Part I)." *Journal of Tropical Pediatrics*, June 1961, 3–15.

———. "Kwashiorkor in Malaya: The Clinical Evidence (Part II)." *Journal of Tropical Pediatrics*, September 1961, 39–48.

———. "The Treatment of Kwashiorkor with Milk and Vegetable Proteins." *British Medical Journal* 2, no. 4788 (1952): 791–96.

———. "Use of Processed Plant Proteins as Human Food." In *Processed Plant Protein Foodstuffs*, edited by Aaron M. Altschul, 205–47. New York: Academic Press, 1958.

Dean, R. F. A., and D. B. Jelliffe. "Diagnosis and Treatment of Protein-Calorie Malnutrition." *Courrier* 10, no. 7 (1960): 429–39.

Dean, R. F. A., and Ruth Schwartz. "The Serum Chemistry in Uncomplicated Kwashiorkor." *British Journal of Nutrition* 7 (1953): 131–47.

Dean, R. F. A., and M. Skinner. "A Note on the Treatment of Kwashiorkor." *Journal of Tropical Pediatrics*, March 1957, 215–16.

Dean, R. F. A., and J. Swanne. "African Child Health: Abbreviated Schedule of Treatment for Severe Kwashiorkor, May 1962." *Journal of Tropical Pediatrics*, 1963, 97–98.

Dean, R. F. A., and B. Weinbren. "Fat Absorption in Chronic Severe Malnutrition in Children." *Lancet* 268, no. 6936 (August 4, 1956): 252.

Defourny, Isabelle. "Operational Innovation in Practice: MSF's Programme Against Malnutrition in Maradi." In Crombé and Jézéquel, *A Not-So Natural Disaster*, 169–87.

Defourny, Isabelle, Andrea Minetti, Géza Harczi, Stéphane Doyon, Susan Sheperd, Milton Tectonidis, Jean-Hervé Bradol, and Michael Golden. "A Large-Scale Distribution of Milk-Based Fortified Spreads: Evidence for a New Approach in Regions with High Burden of Acute Malnutrition." *PLoS ONE* 4, no. 5 (May 6, 2009): 1–7.

Defourny, Isabelle, Gwenola Seroux, Issaley Abdelkader, and Géza Harczi. "Management of Moderate Acute Malnutrition with RUTF in Niger." *Field Exchange* 31 (September 2007): 1–4.

Enserink, Martin. "The Peanut Butter Debate." *Science*, n.s., 322, no. 5898 (October 3, 2008): 36–38.

Gatchell, Valerie, Vivienne Forsythe, and Paul-Rees Thomas. "The Sustainability of Community-Based Therapeutic Care (CTC) in Non-acute Emergency Contexts." In SCN Nutrition Policy Paper No. S21, edited by Claudine Prudhon, André Briend, Zita Weise Prinzo, Bernadette M. E. G. Daelmans, and John B. Mason, *Food and Nutrition Bulletin* 27, no. 3 (September 2006): 6. http://www.who.int/nutrition/publications/severemalnutrition/FNB_0379_5721/en/.

Geber, Marcelle. "The Psycho-Motor Development of African Children in the First Year, and the Influence of Maternal Behavior." *Journal of Social Psychology* 47 (1958): 185–95.

Geber, Marcelle, and R. F. A. Dean. "Gesell Tests on African Children." *Pediatrics*, December 1957, 1055–65.

———. "The Psychological Changes Accompanying Kwashiorkor." *Courrier* 6, no. 1 (January 1956): 3–14.

———. "The State of Development of Newborn African Children." *Lancet*, June 15, 1957, 1216–19.

Golden, Michael H. N., and André Briend. "Treatment of Malnutrition in Refugee Camps." *Lancet* 342, no. 8867 (August 7, 1993): 360.

Gopalan, C. "Kwashiorkor in Uganda and Coonoor." *Journal of Tropical Pediatrics*, March 1956, 206–19.

Holmes, E. G., F. L. Gee, S. B. Asea, S. Bhima, W. K. Chagula, T. R. Evelia, J. W. S. Kasirye et al. "Red Blood Count of East African Students." *East African Medical Journal* 27, no. 9 (1950): 360–70.

Holmes, E. G., M. W. Stanier, Y. B. Semambo, and E. R. Jones, with the assistance of J. Kyobe and a statistical note by F. L. Gee. "An Investigation of Serum Proteins of Africans in Uganda." *Transactions of the Royal Society of Tropical Medicine and Hygiene* 45, no. 3 (1951): 371–82.

Horton, Richard. "Maternal and Child Undernutrition: An Urgent Opportunity." *Lancet* 371 (January 19, 2008): 179.

Horton, Richard, and Selina Lo. "Nutrition: A Quintessential Sustainable Development Goal." *Lancet* 382 (August 3, 2013): 1–2.

Howells, G. R., and R. G. Whitehead. "A System for the Estimation of the Urinary Hydroxyproline Index." *Journal of Medical Laboratory Technology* 24 (1967): 98–102.

Isanaka, Sheila, Nohelly Nombela, Ali Djibo, Marie Poupard, Dominique Van Bekhoven, Valérie Gaboulaud, Philippe J. Guerin, and Rebecca F. Grais. "Effect of Preventive Supplementation with Ready-to-Use Therapeutic Food on the Nutritional Status, Mortality, and Morbidity of Children Aged 6 to 60 Months in Niger: A Cluster Randomized Trial." *Journal of American Medical Association* 301, no. 3 (January 21, 2009): 277–85.

Jelliffe, D. B. "Commerciogenic Malnutrition." *Food Technology* 25 (1971): 153–54.

———. "Custom and Child Health in Buganda: I. Introduction." *Tropical and Geographical Medicine* 15 (1963): 121–23.

———. "Protein-Calorie Malnutrition in Tropical Preschool Children: A Review of Recent Knowledge." *Journal of Pediatrics* 54 (1959): 227–56.

———. "Social Culture and Nutrition: Cultural Blocks and Protein Malnutrition in Early Childhood in Rural West Bengal." *Pediatrics*, July 1957, 128–38.

Jelliffe, D. B., and F. J. Bennett. "Indigenous Medical Systems and Child Health." *Journal of Tropical Pediatrics* 57, no. 2 (August 1960): 248–61.

———. "Nutrition Education in Tropical Child Health Centres (Some Practical Guidelines)." *Courrier* 10, no. 9 (1960): 569–73.

Jelliffe, D. B., F. J. Bennett, E. F. P. Jelliffe. and R. H. R. White. "Ecology of Childhood Disease in the Karamojong of Uganda." *Archives of Environmental Health and Childhood Disease* 9 (1964): 25–36.

Jelliffe, D. B., F. J. Bennett, C. E. Stroud, M. E. Novotnv, H. A. Karrach, L. K. Musoke, and E. F. P. Jelliffe. "Field Survey of the Health of Bachiga Children in the Kayonza District of Kigezi, Uganda." *American Journal of Tropical Medicine and Hygiene* 10 (1961): 435–45.

Jelliffe, D. B., F. J. Bennett, C. E. Stroud, Hebe F. Welbourn, M. C. Williams, and E. F. P. Jelliffe. "The Health of Acholi Children." *Tropical and Geographical Medicine* 15 (1963): 411–12.

Jelliffe, D. B., F. J. Bennett, R. H. R. White, T. R. Cullinan, and E. F. P. Jelliffe. "The Children of the Lugbara: A Study in the Techniques of Paediatric Field Survey in Tropical Africa." *Tropical and Geographical Medicine* 14 (1962): 33–50.

Jelliffe, D. B., and R. F. A. Dean. "Protein-Calorie Malnutrition in Early Childhood (Practical Notes)." *Journal of Tropical Pediatrics*, December 1959, 96–106.

Jelliffe, D. B., and E. F. Patrice Jelliffe. *Human Milk in the Modern World: Psychosocial, Nutritional and Economic Significance.* Oxford: Oxford University Press, 1979.

Jelliffe, D. B., C. Morton, and G. Nansubuga. "*Ettu* Pastes in Infant Feeding in Buganda." *Journal of Tropical Medicine and Hygiene* 65 (1962): 43.

Jelliffe, D. B., B. E. R. Symonds, and E. F. P. Jelliffe. "Childhood Malnutrition and the Technical Development of Tropical Regions." *East African Medical Journal* 37, no. 5 (May 1960): 405–9.

Jones, P. R. M., and R. F. A. Dean. "The Effects of Kwashiorkor on the Development of the Bones and of the Hand." *Journal of Tropical Pediatrics*, September 1956, 51–68.

———. "The Effects of Kwashiorkor on the Development of the Bones and of the Knee." *Journal of Pediatrics*, 1959, 176–84.

King, Maurice, ed. *Medical Care in Developing Countries: A Primer on the Medicine of Poverty and a Symposium from Makerere.* Nairobi, Ken.: Oxford University Press, 1966.

Komrska, Isabelle. "Increasing Access to Ready-to-Use Therapeutic Foods (RUTF)." *Field Exchange* 42 (January 2012): 46.

Krawinkel, Michael. "Kwashiorkor Is Still Not Fully Understood." *Bulletin of the World Health Organization* 81, no. 12 (2003): 910–11.

Latham, Michael C. "International Perspectives on Weaning Foods: The Economic and Other Implications of Bottle Feeding and the Use of Manufactured Weaning Foods." In *Breastfeeding and Food Policy in a Hungry World*, edited by Dana Raphael, 119–27. New York: Academic Press, 1979.

Lelijveld, Natasha, Andrew Seal, Jonathan C. Wells, Jane Kirkby, Charles Opondo, Emmanuel Chimwezi, James Bunn et al. "Chronic Disease Outcomes after Severe Acute Malnutrition in Malawian Children (ChroSAM): A Cohort Study." *Lancet Global Health* 4, no. 9 (September 2016): e654–e662. doi:10.1016/S2214-109X(16)30133-4.

Lutwama, J. S. W. "Uganda: What Children Want to Know about Health." *International Journal of Health Education* 7, no. 2 (1964): 82–90.

McCollum, E. V. *The Newer Knowledge of Nutrition.* New York: The MacMillan Company, 1918.

McLaren, Donald S. "The Great Protein Fiasco." *Lancet* 304 (July 1974): 93–96.

———. "The Great Protein Fiasco Revisited." *Nutrition* 16 (2000): 464–65.

Müller, Olaf, and Michael Krawinkel. "Malnutrition and Health in Developing Countries." *Canadian Medical Association Journal* 173, no. 3 (2005): 279–86.

Musoke, Latimer K. "An Analysis of Admissions to the Paediatric Division, Mulago Hospital in 1959." *Archives of Disease in Childhood* 36 (1961): 305–15.

Muwazi, E. M. K., H. C. Trowell, and J. N. P. Davies. "Congenital Syphilis in Uganda." *East African Medical Journal* 24, no. 4 (April 1947): 152–70.

Namboze, Josephine M. "Mobile Young Children's Clinics in Kasangati Health Centre Defined Area." *Journal of Tropical Pediatrics* 12, S3 (1966): 75–77.

———. "A Study of Births and Deaths in the Defined Area of Kasangati Health Centre in the Year 1967." *Journal of Tropical Pediatrics*, September 1969, 99–108.

———. "Weaning Practices in Buganda." *Tropical and Geographical Medicine* 19 (1967): 154–56.

Neufeld, Lynnette M. "Ready-to-Use Therapeutic Food for the Prevention of Wasting in Children." *Journal of American Medical Association* 301, no. 3 (January 21, 2009): 327–28.

Orr, John Boyd. *Food, Health and Income: Report on a Survey of Adequacy of Diet in Relation to Income*. London: Macmillan & Co., 1936.

Orr, John Boyd, and J. L. Gilks. "Studies of Nutrition: The Physique and Health of Two African Tribes." *Medical Research Council Annual Report*, Special Report Series 155 (1931): 5–33.

Osorio, Snezana Nena. "Reconsidering Kwashiorkor." *Topical Clinical Nutrition* 26, no. 1 (2011): 10–13.

Robinson, D. C. "The Nutrition Rehabilitation Unit at Mulago Hospital, Kampala: Further Development and Evaluation 1967-69." In Stanfield, "Malnutrition in Uganda," 35–42.

Rowland, M. G. M., and J. P. K. McCollum. "Malnutrition and Gastroenteritis in the Gambia." *Transactions of the Royal Society of Tropical Medicine and Hygiene* 71, no. 3 (1977): 199–203.

Rutishauser, I. H. E. "Custom and Child Health in Buganda: IV. Food and Nutrition." *Tropical and Geographical Medicine* 15 (1963): 130–47.

———. "Heights and Weights of Middle Class Baganda Children." *Lancet*, September 18, 1965, 565–67.

———. "Statistics of Malnutrition in Early Childhood (with Reference to Uganda)." In Stanfield, "Malnutrition in Uganda," 11–16.

Rutishauser, I. H. E., and R. G. Whitehead. "Field Evaluation of Two Biochemical Tests which May Reflect Nutritional Status in Three Areas of Uganda." *British Journal of Nutrition* 23, no. 1 (1969): 1–13.

Schneideman, I., F. J. Bennett, and I. H. E. Rutishauser. "The Nutrition Rehabilitation Unit at Mulago Hospital-Kampala: Development and Evaluation, 1965–67." In Stanfield, "Malnutrition in Uganda," 25–34.

Scrimshaw, Nevin S. "A Look at the Incaparina Experience in Guatemala: The Background and History of Incaparina." *Food and Nutrition Bulletin* 2, no. 2 (1980): 1–2.

Scrimshaw, Nevin S., and Moisés Béhar. "Protein Malnutrition in Young Children." *Science* 133, no. 3470 (June 1961): 2039–47.

Scwhartz, Ruth, and R. F. A. Dean. "The Serum Lipids in Kwashiorkor: I. Neutral Fat, Phospholipids and Cholesterol." *Journal of Tropical Pediatrics* 3, no. 1 (June 1957): 23-31.

Stanfield, J. P. "Introduction." In Stanfield, "Malnutrition in Uganda," 3–4.

———. "The Luteete Family Health Centre: Nutrition Rehabilitation in a Comprehensive Rural Development Strategy." In Stanfield, "Malnutrition in Uganda," 67–74.

———. "Nutrition Education in the Context of Early Childhood Malnutrition in Low-Resource Communities." *Proceedings of the Nutrition Society* 35 (1976): 131–38.

———., ed. "Recent Approaches to Malnutrition in Uganda." Monograph No. 13 *Journal of Tropical Pediatrics and Environmental Child Health* 17, no. 1 (March 1971).

Stanier, M., and M. D. Thompson. "The Serum Protein Levels of Newborn African Infants." *Archives of Disease in Childhood* 29, no. 144 (1954): 110–12.

Stannus, H. S. "A Nutritional Disease of Childhood Associated with a Maize Diet—and Pellagra." *Archives of Disease in Childhood* 9, no. 50 (1934): 115–18.

Thompson, M. D., and H. C. Trowell. "Pancreatic Enzyme Activity in the Duodenal Contents of Children with a Type of Kwashiorkor." *Lancet*, 1952, 1031–35.

Trehan, Indi, Hayley S. Goldbach, Lacey N. LaGrone, Guthrie J. Meuli, Richard J. Wang, Kenneth M. Maleta, and Mark J. Manary. "Antibiotics as Part of the Management of Severe Acute Malnutrition." *New England Journal of Medicine* 368, no. 5 (January 31, 2013): 425–35.

Trowell, H. C. "Food, Protein and Kwashiorkor: A Presidential Address, 1956." *Uganda Journal* 21 (1957): 81–90.

———. "A Note on Infantile Pellagra." *Transactions of the Royal Society of Medicine and Hygiene* 35, no. 1 (1941): 18–20.

———. "Pellagra in African Children." *Archives of Disease in Childhood* 12 (1937): 193–212.

Trowell, H. C., J. N. P. Davies, and R. F. A. Dean. *Kwashiorkor.* London: Academic Press, 1982. First published in 1954.

———. "Kwashiorkor: II. Clinical Picture, Pathology, and Differential Diagnosis." *British Medical Journal*, October 11, 1952, 798–801.

Trowell, H. C., and D. B. Jelliffe, eds. *Diseases of Children in the Subtropics and Tropics.* London: Edward Arnold, 1958.

Trowell, H. C., and E. M. K. Muwazi. "A Contribution to the Study of Malnutrition in Central Africa: A Syndrome of Malignant Malnutrition." *Transactions of the Royal Society of Tropical Medicine and Hygiene* 39, no. 3 (1945): 229–43.

———. "Severe and Prolonged Underfeeding in African Children (the Kwashiorkor Syndrome of Malignant Malnutrition)." *Archives of Disease in Childhood* 20 (1945): 110–16.

Watts, E. R., and D. G. R. Belshaw. "The Small Holding at Luteete Family Health Centre." In Stanfield, "Malnutrition in Uganda," 83–84.

Welbourn, Hebe F. "The Danger Period during Weaning." *Journal of Tropical Pediatrics*, June 1955, 34–46.

———. "The Danger Period during Weaning (Part II)." *Journal of Tropical Pediatrics*, September 1955, 98–111.

———. "Child Welfare in Mengo District, Uganda." *Journal of Tropical Pediatrics*, June 1956, 24–31.

———. "Bottle Feeding: A Problem of Modern Civilization." *Journal of Tropical Pediatrics*, March 1958, 157–66.

———. "Backgrounds and Follow-Up of Children with Kwashiorkor." *Journal of Tropical Pediatrics*, December 1959, 84–95.

———. "Weaning among the Baganda." *Journal of Tropical Pediatrics and African Child Health* 9 (June 1963): 14–24.

———. "Custom and Child Health in Buganda: II. Methods of Child Rearing." *Tropical and Geographical Medicine* 15 (1963): 124–33.

———. "Custom and Child Health in Buganda: III. Pregnancy and Childbirth." *Tropical and Geographical Medicine* 15 (1963): 134–37.

Welbourn, Hebe F., and Grace De Beer. "Trial of a Kit for Artificial Feeding in Tropical Village Homes." *Journal of Tropical Medicine and Hygiene* 67, no. 155 (1964): 155–59.

Whitehead, R. G. "Amino Acid Metabolism in Kwashiorkor, I. Metabolism of Histidine and Imidazole Derivatives." *Clinical Science* 26 (1964): 271–78.

———. "The Assessment of Nutritional Status in Protein-Malnourished Children." *Proceedings of the Nutrition Society* 28 (1969): 1–16.

———. "Biochemical Tests in Differential Diagnosis of Protein and Calorie Deficiencies." *Archives of Diseases in Childhood* 42 (1967): 479–84.

———. "Hydroxyproline Creatinine Ratio as an Index of Nutritional Status and Rate of Growth." *Lancet*, September 1965, 567–70.

———. "Kwashiorkor in Uganda." In *The Contribution of Nutrition to Human and Animal Health*, edited by E. M. Widdowson and J. C. Mathers, 303–13. Cambridge: Cambridge University Press, 1992.

———. "Rapid Determination of Some Plasma Aminoacids in Subclinical Kwashiorkor." *Lancet*, February 1964, 250–52.

———. "An Unidentified Compound in the Serum of Children with Kwashiorkor (Protein-Calorie Malnutrition)." *Nature* 204, no. 4956 (October 1964): 389.

Whitehead, R. G., W. A. Coward, P. G. Lunn, and Ingrid Rutishauser. "A Comparison of the Pathogenesis of Protein-Energy Malnutrition in Uganda and the Gambia." *Transactions of the Royal Society of Tropical Medicine and Hygiene* 71, no. 3 (1977): 189–95.

Whitehead, R. G., and R. F. A. Dean. "Serum Amino Acids in Kwashiorkor: I. Relationship to Clinical Condition." *American Journal of Clinical Nutrition* 14, no. 6 (1964): 313–19.

———. "Serum Amino Acids in Kwashiorkor: II. An Abbreviated Method of Estimation and Its Application." *American Journal of Clinical Nutrition* 14, no. 6 (1964): 320–30.

Whitehead, R. G., and Celia E. Matthew. "The Analysis of Urine of Children Suffering from Kwashiorkor." *East African Medical Journal* 37, no. 5 (May 1960): 384–90.

Whitehead, R. G., and T. R. Milburn. "Amino Acid Metabolism in Kwashiorkor, II. Metabolism of Phenylalanine and Tyrosine." *Clinical Science* 26 (1964): 279–89.

Williams, C. D. "A Nutritional Disease of Childhood Associated with a Maize Diet." *Archives of Disease in Childhood* 8 (1933): 423–33.

———. "Kwashiorkor: A Nutritional Disease of Children Associated with a Maize Diet." *Lancet*, November 16, 1935, 1151–52.

———. "The Organisation of Child Health Services in Developing Countries." *Journal of Tropical Pediatrics*, June 1955, 3–7.

———. "Trends in Social Pediatrics." *Journal of Tropical Pediatrics*, December 1961, 83–86.

General Works

Adichie, Chimamanda Ngozi. *Half of a Yellow Sun*. New York: Alfred A. Knopf, 2006.

Ainsworth, Mary D. Salter. *Infancy in Uganda: Infant Care and the Growth of Love*. Baltimore: Johns Hopkins Press, 1967.

Allman, Jean, Susan Geiger, and Nakanyike Musisi, eds. *Women in African Colonial Encounters*. Bloomington: Indiana University Press, 2002.

Anderson, Warwick. "Immunities of Empire: Race, Disease, and the New Tropical Medicine, 1900-1920." *Bulletin of the History of Medicine* 70, no. 1 (1996): 94–118.

———. "Making Global Health History: The Postcolonial Worldliness of Biomedicine." *Social History of Medicine* 27, no. 2 (2014): 372–84.

Apter, David E. *The Political Kingdom in Uganda: A Study in Bureaucratic Nationalism*. New Jersey: Princeton University Press, 1961.

Arnold, David. "The Place of 'The Tropics' in Western Medical Ideas since 1750." *Tropical Medicine and International Health* 2, no. 4 (1997): 303–13.

———. *Imperial Medicine and Indigenous Societies*. New York: Manchester University Press, 1998.

Asiimwe, Godfrey B. "From Monopoly Marketing to Coffee *Magendo*: Responses to Policy Recklessness and Extraction in Uganda, 1971–79." *Journal of Eastern African Studies* 7, no. 1 (February 2013): 104–24.

Austoker, J., and L. Bryder. *Historical Perspectives on the Role of the MRC: Essays in the History of the Medical Research Council of the United Kingdom and Its Predecessor the Medical Research Committee, 1913–1953*. New York: Oxford University Press, 1989.

Basudde, Elvis. "Uganda's Forgotten Music Heroes." *New Vision*, Tuesday, January 22, 2008.

Beinart, Jennifer. "The Inner World of Imperial Sleeping Sickness: The MRC and Research in Tropical Medicine." In Austoker and Bryder, *Historical Perspectives*, 109–26.

Bell, Heather. *Frontiers of Medicine in the Anglo-Egyptian Sudan, 1899–1940.* Oxford: Clarendon Press, 1999.

Biehl, João Guilherme. *Will to Live: Aids Therapies and the Politics of Survival.* Princeton: Princeton University Press, 2007.

Biehl, João Guilherme, and Adriana Petryna, eds. *When People Come First: Critical Studies in Global Health.* Princeton: Princeton University Press, 2013.

Birungi, Harriet. "Injections and Self-Help: Risk and Trust in Ugandan Health Care." *Social Science & Medicine* 47, no. 10 (November 1998): 1455–62.

Bivins, Roberta. "Coming 'Home' to (post)Colonial Medicine: Treating Tropical Bodies in Post-war Britain." *Social History of Medicine* 26, no. 1 (2013): 1–20.

Black, Maggie. *Children First: The Story of UNICEF, Past and Present.* Oxford: Oxford University Press, 1996.

———. *The Children and the Nations: The Story of Unicef.* New York: Unicef, 1986.

Bowles, B. D. "Economic Anti-colonialism and British Reaction in Uganda, 1936–1955." *Canadian Journal of African Studies* 9, no. 1 (1975): 51–60.

Boyd, Lydia. *Preaching Prevention: Born-Again Christianity and the Moral Politics of AIDS in Uganda.* Athens: Ohio University Press, 2015.

Brantley, Cynthia. "Kikuyu-Maasai Nutrition and Colonial Science: The Orr and Gilks Study in Late 1920s Kenya Revisited." *International Journal of African Historical Studies* 30 (1999): 49–86.

———. *Feeding Families: African Realities and British Ideas of Nutrition and Development in Early Colonial Africa.* Portsmouth: Heinemann, 2002.

Carpenter, K. J. *Protein and Energy: A Study of Changing Ideas in Nutrition.* New York: Cambridge University Press, 1994.

Chesswas, J. D. *The Essentials of Luganda.* 1954; East African Literature Bureau Edition, Nairobi: Oxford University Press, 2002.

Clarke, Adele E., Janet K. Shim, Laura Mamo, Jennifer Ruth Fosket, and Jennifer R. Fishman. "Biomedicalization: Technoscientific Transformations of Health, Illness, and U.S. Biomedicine." *American Sociological Review* 68, no. 2 (2003): 161–94.

Cook, Albert R. "The Medical History of Uganda: Part II." *East African Medical Journal* 13 (1937): 99–110.

———. *Uganda Memories (1897–1940).* Kampala: The Uganda Society, 1945.

Cooper, Anderson. "'Miracle' Food Saves Lives." *60 Minutes* video. Posted June 22, 2008. http://www.cbsnews.com/videos/miracle-food-saves-lives/.

Cooper, Barbara M. "Chronic Malnutrition and the Trope of the Bad Mother." In Crombé and Jézéquel, *A Not-So Natural Disaster*, 147–68.

———. *Evangelical Christians in the Muslim Sahel.* Bloomington: Indiana University Press, 2006.

———. *Marriage in Maradi: Gender and Culture in a Hausa Society in Niger, 1900–1989.* Portsmouth: Heinemann, 1997.

Cooper, Frederick. *Colonialism in Question: Theory, Knowledge, History.* Berkeley: University of California Press, 2005.

———. "Conflict and Connection: Rethinking Colonial African History." *American Historical Review* 99, no. 5 (1994): 1516–45.

———. *Decolonization and African Society: The Labor Question in French and British Africa.* Cambridge: Cambridge University Press, 1996.

Cooper, Frederick, and Ann Laura Stoler, eds. *Tensions of Empire: Colonial Cultures in a Bourgeois World.* Berkeley: University of California Press, 1997.

Crane, Johanna Tayloe. *Scrambling for Africa: AIDS, Expertise, and the Rise of American Global Health Science.* Ithaca: Cornell University Press, 2013.

Crombé, Xavier, and Jean-Hervé Jézéquel, eds. *A Not-So Natural Disaster: Niger 2005.* New York: Columbia University Press, 2009.

Curtin, Philip D. "Medical Knowledge and Urban Planning in Colonial Tropical Africa." In Feierman and Janzen, *Social Basis of Health,* 235–55.

Dawson, Marc H. "The 1920s Anti-yaws Campaign and Colonial Medical Policy in Kenya" *International Journal of African Historical Studies* 20, no. 3 (1987): 417–35.

Decker, Alicia. *In Idi Amin's Shadow: Women, Gender and Militarism in Uganda.* Athens: Ohio University Press, 2014.

"Derrick B. Jelliffe, Public Health: Los Angeles." In *University of California: In Memoriam, 1993,* edited David Krogh. Berkeley: University of California, 1993. http://content.cdlib.org/view?docId=hb0h4n99rb&doc.view=frames&chunk.id=div00034&toc.depth=1&toc.id=&brand=calisphere.

Dimock, Elizabeth. "Women's Leadership Roles in the Early Protestant Church in Uganda: Continuity with the Old Order." *Australian Review of African Studies* 25 (2003): 8–22.

Dilger, Hansjörg, Abdoulaye Kane, and Stacey Ann Langwick, eds. *Medicine, Mobility, and Power in Global Africa: Transnational Health and Healing.* Bloomington: Indiana University Press, 2012.

Dobbing, John, ed. *Infant Feeding: Anatomy of a Controversy, 1973–1984.* New York: Springer-Verlag, 1988.

Dodge, Cole P., and Paul D. Wiebe, eds. *Crisis in Uganda: The Breakdown of Health Services.* New York: Pergamon Press, 1985.

Doyle, Shane. *Before HIV: Sexuality, Fertility and Mortality in East Africa, 1900–1980.* Oxford: Oxford University Press, 2013.

———. *Crisis and Decline in Bunyoro: Population and Environment in Western Uganda, 1860–1955.* Athens: Ohio University Press, 2006.

———."Population Decline and Delayed Recovery in Bunyoro, 1860–1960." *Journal of African History* 41 (2000): 429–58.

Echenberg, Myron. *Black Death, White Medicine: Bubonic Plague and the Politics of Public Health in Colonial Senegal, 1914–1945.* Portsmouth, NH: Heinemann, 2002.

Ellis, Stephen. "Writing Histories of Contemporary Africa." *Journal of African History* 43 (2002): 1–26.

Ehrlich, Cyril. "The Uganda Economy, 1903–1945." In *History of East Africa*, edited by Vincent Harlow and E. M. Chilver, 395–75. Oxford: Oxford University Press, 1965.

Fairhead, James, Melissa Leach, and Mary Small. "Where Techno-Science Meets Poverty: Medical Research and the Economy of Blood in The Gambia, West Africa." *Social Science and Medicine* 63 (2006): 1109–20.

Fallers, L. A., ed. *The King's Men: Leadership and Status in Buganda on the Eve of Independence*. London: Oxford University Press, 1964.

Fallers, L. A., assisted by F. K. Kamoga and S. B. K. Musoke. "Social Stratification in Traditional Buganda." In Fallers, *The King's Men*, 105–6.

Fallers, L. A., and S. B. K. Musoke. "Social Mobility, Traditional and Modern." In Fallers, *The King's Men*, 193.

Fassin, Didier. *When Bodies Remember: Experiences and Politics of AIDS in South Africa*. Berkeley: University of California Press, 2007.

Feierman, Steven. "Africa in History: The End of Universal Narratives." In *After Colonialism: Imperial Histories and Postcolonial Displacements*, edited by Gyan Prakash, 40–65. Princeton: Princeton University Press, 1995.

———. "African Histories and the Dissolution of World History." In *Africa and the Disciplines: The Contributions of Research in Africa to the Social Sciences and Humanities*, edited by Robert H. Bates, V. Y. Mudimbe, and Jean O'Barr, 167–79. Chicago: University of Chicago Press, 1993.

———. "Change in African Therapeutic Systems." In "The Social History of Disease and Medicine in Africa." Special issue, *Social Science and Medicine* 13B (1979): 277–84.

———. "Colonizers, Scholars, and the Creation of Invisible Histories." In *Beyond the Cultural Turn: New Directions in the Study of Society and Culture*, edited by Victoria E. Bonnell and Lynn Hunt, 182–215. Berkeley: University of California Press, 1999.

———. "Explanation and Uncertainty in the Medical World of Ghaambo." *Bulletin of the History of Medicine* 74 (2000): 317–44.

———. *Peasant Intellectuals: Anthropology and History in Tanzania*. Madison: University of Wisconsin Press, 1990.

———. "Struggles for Control: The Social Roots of Health and Healing in Modern Africa." *African Studies Review* 28, no. 2/3 (1985): 73–147.

Feierman, Steven, and John M. Janzen, eds. *The Social Basis of Health and Healing in Africa*. Berkeley: University of California Press, 1992.

Ferguson, James. *Expectations of Modernity: Myths and Meanings of Urban Life on the Zambian Copperbelt*. Berkeley: University of California Press, 1999.

———. *Global Shadows: Africa in the Neoliberal World Order*. Durham, NC: Duke University Press, 2006.

Ford, John. *The Role of the Trypansomiasis in African Ecology: A Study of the Tsetse Fly Problem*. Oxford: Clarendon Press, 1971.

Fullwiley, Duana. *The Enculturated Gene: Sickle Cell Health Politics and Biological Difference in West Africa*. Princeton: Princeton University Press, 2011.

Gaitskell, Deborah. "Housewives, Maids or Mothers: Some Contradictions of Domesticity for Christian Women in Johannesburg, 1903–1939." *Journal of African History* 24 (1983): 241–56

Geach, Hugh. "The Baby Food Tragedy." *New Internationalist*, no. 006 (August 1973): 8–12, 23.

Geissler, Wenzel P. "'Kachinja Are Coming!': Encounters around Medical Research Work in a Kenyan Village." *Journal of the International African Institute* 75, no. 2 (2005): 173–202.

———, ed. *Para-States and Medical Science: Making African Global Health*. Durham, NC: Duke University Press, 2015.

Geissler, Wenzel P., Ann Kelly, Babatunde Imoukhuede, and Robert Pool. "'He Is Now Like A Brother, I Can Even Give Him Some Blood'– Relational Ethics and Material Exchanges in a Malaria Vaccine 'Trial Community' in the Gambia." *Social Science and Medicine* 67 (2008): 696–707.

Geissler, Wenzel P., and Catherine Molyneux, eds. *Evidence, Ethos and Experiment: The Anthropology and History of Medical Research in Africa*. New York: Berghahn, 2011.

George, Abosede A. *Making Modern Girls: A History of Girlhood, Labor, and Social Development in Colonial Lagos*. Athens: Ohio University Press, 2014.

Giblin, James. "Trypanosomiasis Control in African History: An Evaded Issue?" *Journal of African History* 31 (1990): 59–80.

Giles-Vernick, Tamara. *Cutting the Vines of the Past: Environmental Histories of the Central African Rain Forest*. Charlottesville: University Press of Virginia, 2002.

Giles-Vernick, Tamara, and James L. A. Webb Jr., eds. *Global Health in Africa: Historical Perspectives on Disease Control*. Perspectives on Global Health. Athens: Ohio University Press, 2013.

Good, Charles M. *The Steamer Parish: The Rise and Fall of Missionary Medicine on an African Frontier*. Chicago: University of Chicago Press, 2004.

Graboyes, Melissa. *The Experiment Must Continue: Medical Research and Ethics in East Africa, 1940–2014*. Perspectives on Global Health. Athens: Ohio University Press, 2015.

———. "Fines, Orders, Fear . . . and Consent?: Medical Research in East Africa, c. 1950s." *Developing World Bioethics* 10, no. 1 (2010): 34–41.

Gregg, Alan. "Henry E. Sigerist: His Impact on American Medicine." *Bulletin of the History of Medicine* 22 (1948): 32–34.

Hamilton, Carolyn. *Terrific Majesty: The Powers of Shaka Zulu and the Limits of Historical Invention*. Cambridge, MA: Harvard University Press, 1998.

Hamlin, Christopher. "Public Health." In *The Oxford Handbook of the History of Medicine*, edited by Mark Jackson, 411–28. Oxford: Oxford University Press, 2011.

Hammonds, Rachel, and Gorik Ooms. "World Bank Policies and the Obligation of Its Members to Respect, Protect and Fulfill the Right to Health." *Health and Human Rights* 8, no. 1 (January 1, 2004): 26–60.

Hansen, Holger Bernt. "Uganda in the 1970s: A Decade of Paradoxes and Ambiguities." *Journal of Eastern African Studies* 7, no. 1 (February 2013): 87–94.

Hansen, Holger Bernt, and Michael Twaddle, eds. *Changing Uganda: The Dilemmas of Structural Adjustment & Revolutionary Change*. Athens: Ohio University Press, 1991.

Hansen, Karen Tranenberg, ed. *African Encounters with Domesticity*. New Brunswick, NJ: Rutgers University Press, 1992.

Hanson, Holly. *Landed Obligation: The Practice of Power in Buganda*. Portsmouth: Heinemann, 2003.

———. "Queen Mothers and Good Government in Buganda: The Loss of Women's Political Power in Nineteenth-Century East Africa." In Allman, Geiger, and Musisi, *Women in African Colonial Encounters*, 219–36.

———. "Stolen People and Autonomous Chiefs in Nineteenth-Century Buganda: The Social Consequences of Non-free Followers." In *Slavery in the Great Lakes Region of East Africa*, edited by Henri Médard and Shane Doyle, 161–73. Athens: Ohio University Press, 2007.

Harrison, Mark. *Climates and Constitutions: Health, Race, Environment and British Imperialism, 1600–1850*. London: Oxford University Press, 1999.

———. "'The Tender Frame of Man': Disease, Climate, and Racial Difference in India and the West Indies." *Bulletin of the History of Medicine* 70, no. 1 (1996): 68–93.

Hattersley, C. W. *The Baganda at Home: With One Hundred Pictures of Life and Work in Uganda*. London: The Religious Tract Society, 1908.

Himbara, David, and Dawood Sultan. "Reconstructing the Ugandan State and Economy: The Challenge of an International Bantustan." *Review of African Political Economy* 22, no. 63 (March 1, 1995): 85–93.

Hodge, Joseph Morgan. *Triumph of the Expert: Agrarian Doctrines of Development and the Legacies of British Colonialism*. Athens: Ohio University Press, 2007.

Hoppe, Kirk Arden. "Lords of the Fly: Colonial Visions and Revisions of African Sleeping-Sickness Environments on Ugandan Lake Victoria, 1906–61." *Africa* 67, no. 1 (1997): 86–105.

———. *Lords of the Fly: Sleeping Sickness Control in British East Africa, 1900–1960*. Westport, CT: Praeger, 2003.

Hunt, Nancy Rose. "'Le Bebe en Brusse': European Women, African Birth Spacing and Colonial Intervention in Breast Feeding in the Belgian Congo." *International Journal of African Historical Studies* 21, no. 3 (1988): 401–32.

———. *A Colonial Lexicon: Of Birth Ritual, Medicalization, and Mobility in the Congo*. Durham, NC: Duke University Press, 1999.

———. *A Nervous State: Violence, Remedies, and Reverie in Colonial Congo*. Durham, NC: Duke University Press, 2016.

Iliffe, John. *The African Aids Epidemic: A History*. Athens: Ohio University Press, 2006.

———. *East African Doctors: A History of the Modern Profession*. 1998; Cambridge edition, Kampala, Uga.: Fountain Publishers, 2002.

Ivers, Louise C., Kimberly A. Cullen, Kenneth A. Freedberg, Steven Block, Jennifer Coates, and Patrick Webb. "HIV/AIDS, Undernutrition, and Food Insecurity." *Clinical Infectious Diseases* 49, no. 7 (October 1, 2009): 1096–98.

Janzen, John M. *Ngoma Discourses of Healing in Central and Southern Africa*. Berkeley: University of California Press, 1992.

———. *The Quest for Therapy in Lower Zaire*. Berkeley: University of California Press, 1978.

Janzen, John M., and Steven Feierman. "Introduction." In "The Social History of Disease and Medicine in Africa," special issue, *Social Science and Medicine* 13B (1979): 239–43.

Johnston, Alastair. "The Luwero Triangle: Emergency Operations in Luwero, Mubende and Mpigi Districts." In Dodge and Wiebe, *Crisis in Uganda*, 97–106.

Kagwa, Apolo. *The Customs of the Baganda*. Edited by May Mandelbaum. Translated by Ernest B. Kalibala. New York: Columbia University Press, 1934.

Karugire, Samwiri Rubaraza. *Roots of Instability in Uganda*. Kampala, Uga.: Fountain Publishers, 1996.

Kasfir, Nelson. "Guerrillas and Civilian Participation: The National Resistance Army in Uganda, 1981–86." *Journal of Modern African Studies* 43, no. 2 (June 16, 2005): 271–96.

Kasozi, A. B. K. *The Social Origins of Violence in Uganda, 1964–1985*. Montreal: McGill-Queen's University Press, 1994.

Kavuma, Paulo. *Crisis in Buganda, 1953–55*. London: Collings, 1979.

Kayemba, Henry. *A State of Blood: The Inside Story of Idi Amin*. New York: Ace Books, 1977.

Kjekshus, Helge. *Ecology Control and Economic Development in East African History: The Case of Tanganyika, 1850–1950*. London: Heinemann, 1977.

Kiiza, Julius, Godfrey Asiimwe, and David Kibikyo. "Understanding Economic and Institutional Reforms in Uganda." In *Understanding Economic Reforms in Africa: A Tale of Seven Nations*, edited by Joseph Mensah, 57–94. New York: Palgrave Macmillan, 2007.

Kitching, A. L., and G. R. Blackledge. *A Luganda-English and English-Luganda Dictionary*. London: Society for Promoting Christian Knowledge, 1925.

Kodesh, Neil. *Beyond the Royal Gaze: Clanship and Public Healing in Buganda*. Charlottesville: University of Virginia Press, 2010.

———. "History from the Healer's Shrine: Genre, Historical Imagination, and Early Ganda History." *Comparative Studies in Society and History* 49, no. 3 (2007): 527–52.

Kuhanen, Jan. *Poverty, Health and Reproduction in Early Colonial Uganda*. Joensuu, Fin.: University of Joensuu, 2005.

Kyaddondo, David, and Susan Reynolds Whyte. "Working in a Decentralized System: A Threat to Health Workers' Respect and Survival in Uganda." *International Journal of Health Planning and Management* 18, no. 4 (October 1, 2003): 329–42.

Kyomuhendo, Grace Bantebya, and Marjorie Keniston McIntosh. *Women, Work and Domestic Virtue in Uganda, 1900–2003*. Athens: Ohio University Press, 2006.

Langwick, Stacey Ann. *Bodies, Politics, and African Healing the Matter of Maladies in Tanzania*. Bloomington: Indiana University Press, 2011.

Lateef, K. Sarwar. "Structural Adjustment in Uganda: The Initial Experience." In Hansen and Twaddle, *Changing Uganda*, 20–42.

Livingston, Julie. *Debility and the Moral Imagination in Botswana*. Bloomington: Indiana University Press, 2005.

———. *Improvising Medicine: An African Oncology Ward in an Emerging Cancer Epidemic*. Durham, NC: Duke University Press, 2012.

Longhurst, Richard. "Famines, Food, and Nutrition: Issues and Opportunities for Policy and Research." *Food and Nutrition Bulletin* 9, no. 1 (1987). http://archive.unu.edu/unupress/food/8F091e/8F091E05.htm#Famines, food, and nutrition: issues and opportunities for policy.

Lyons, Maryinez. *The Colonial Disease: A Social History of Sleeping Sickness in Northern Zaire, 1900–1940*. Cambridge: Cambridge University Press, 1992.

Low, D. A. *Buganda in Modern History*. Berkeley: University of California Press, 1971.

———, ed. *The Mind of Buganda: Documents of the Modern History of an African Kingdom*. Berkeley: University of California Press, 1971.

Low, D. A., and R. C. Pratt. *Buganda and British Overrule: Two Studies*. London: Oxford University Press, 1970.

Mair, Lucy. *An African People in the Twentieth Century*. 1934; reprint, New York: Russell & Russell, 1965.

Mandala, Elias C. *The End of Chidyerano: A History of Food and Everyday Life in Malawi, 1860–2004*. Portsmouth: Heinemann, 2005.

Mann, Gregory. "An Africanist's Apostasy: On Luise White's 'Speaking with Vampires'." *International Journal of African Historical Studies* 41, no. 1 (2008): 117–21.

———. *From Empires to NGOs in the West African Sahel: The Road to Nongovernmentality*. African Studies Series 129. New York: Cambridge University Press, 2015.

———. *Native Sons: West African Veterans and France in the Twentieth Century*. Durham, NC: Duke University Press, 2006.

Marks, Shula. "Doctors and the State: George Gale and South Africa's Experiment in Social Medicine." In *Science and Society in Southern Africa*, edited by Saul Dubow, 188–211. New York: Manchester University Press, 2000.

———. "What Is Colonial about Colonial Medicine? And What Has Happened to Imperialism and Health." *Society for the Social History of Medicine* 10, no. 2 (1997): 205–19.

Marks, Shula, and Neil Andersson. "Industrialization, Rural Health, and the 1944 National Health Services Commission in South Africa." In Feierman and Janzen, *Social Basis of Health*, 131–61.

Maturo, Antonio. "Medicalization: Current Concept and Future Directions in a Bionic Society." *Mens Sana Monographs* 10, no. 1 (2012): 122–33.

Miller, Joseph C., ed. *The African Past Speaks: Essays on Oral Tradition and History*. Hamden, CT: Dawson-Archon, 1980.

Moore, Henrietta L., and Megan Vaughan. *Cutting Down Trees: Gender, Nutrition, and Agricultural Change in the Northern Province of Zambia, 1890–1990*. Portsmouth: Heinemann, 1994.

Moran, Amy Thomas. "A Salvage Ethnography of the Guinea Worm: Witchcraft, Oracles and Magic in a Disease Eradication Program." In Biehl and Petryna, *When People Come First*, 207–39.

Mugyenyi, Joshua B. "IMF Conditionality and Structural Adjustment Under the National Resistance Movement." In Hansen and Twaddle, *Changing Uganda*, 61–77.

Mukwaya, A. B. *Land Tenure in Buganda: Present Day Tendencies*. Kampala, Uga.: Eagle Press, 1953.

Muller, Mike. *The Baby Killer: A War on Want Investigation into the Promotion and Sale of Powdered Baby Milks in the Third World*. London: War on Want, 1974.

Musisi, Nakanyike B. "The Politics of Perception or Perception as Politics? Colonial and Missionary Representations of Baganda Women, 1900–1945." In Allman, Geiger, and Musisi, *Women in African Colonial Histories*, 95–115.

———. "Women, 'Elite Polygyny,' and Buganda State Formation." *Signs: Journal of Women in Culture and Society* 16 (1991): 757–86.

Nabiruma, Diana. "Mulago's Mwanamugimu Gets a Facelift," *Observer*, June 8, 2001. http://allafrica.com/stories/201106090983.html.

Nannyonga-Tamusuza, Sylvia. "Constructing the Popular: Challenges of Archiving Ugandan 'Popular' Music." *Current Writing: Text and Reception in Southern Africa* 18, no. 2 (2006): 33–52.

Neill, Deborah. "Finding the 'Ideal Diet': Nutrition, Culture, and Dietary Practices in France and French Equatorial Africa, c. 1890s to 1920s." *Food and Foodways* 17, no. 1 (2009): 1–28.

Nguyen, Vinh-Kim. *The Republic of Therapy: Triage and Sovereignty in West Africa's Time of AIDS*. Durham, NC: Duke University Press, 2010.

———. "Treating to Prevent HIV: Population Trials and Experimental Societies." In Geissler, *Para-States and Medical Science*, 47–77.

Nuttall, Sarah. *Entanglement: Literary and Cultural Reflections on Post-apartheid*. Johannesburg: Wits University Press, 2009.

Odonga, Alexander M. *The First Fifty Years of Makerere University Medical School and the Foundation of Scientific Medical Education in East Africa*. Kisubi, Uga.: Marianum Press, 1989.

Okuonzi, Sam Agatre, and Joanna Macrae. "Whose Policy Is It Anyway? International and National Influences on Health Policy Development in Uganda." *Health Policy and Planning* 10, no. 2 (1995): 122–32.

Osseo-Asare, Abena Dove. *Bitter Roots: The Search for Healing Plants in Africa*. Chicago: University of Chicago Press, 2014.

Packard, Randall. *A History of Global Health: Interventions into the Lives of Other Peoples*. Baltimore: Johns Hopkins University Press, 2016.

———. "Malaria Dreams: Postwar Visions of Health and Development in the Third World." *Medical Anthropology* 17 (1997): 279–96.

———. "Visions of Postwar Health and Development and Their Impact on Public Health Interventions in the Developing World." In *International Development and the Social Sciences: Essays on the Politics of Knowledge*, edited by Frederick Cooper and Randall Packard, 93–115. Berkeley: University of California Press, 1997.

———. *White Plague, Black Labor: Tuberculosis and the Political Economy of Health and Disease in South Africa*. Berkeley: University of California Press, 1989.

Peterson, Derek R. "Morality Plays: Marriage, Church Courts, and Colonial Agency in Central Tanganyika, ca. 1876-1928." *American Historical Review* 111, no. 4 (October 2006): 983–1010.

Petryna, Adriana. *When Experiments Travel: Clinical Trials and the Global Search for Human Subjects*. Princeton: Princeton University Press, 2009.

Petty, Celia. "Primary Research and Public Health: The Prioritization of Nutrition Research in Inter-war Britain," in Austoker and Bryder, *Historical Perspectives*, 83–108.

Pfeiffer, James. "International NGOs and Primary Health Care in Mozambique: The Need for a New Model of Collaboration." *Social Science & Medicine* 56, no. 4 (February 2003): 725–38.

———. "The Struggle for a Public Sector: PEPFAR in Mozambique." In Biehl and Petryna, *When People Come First*, 166–81.

Pfeiffer, James, and Rachel Chapman. "Anthropological Perspectives on Structural Adjustment and Public Health." *Annual Review of Anthropology* 39 (January 1, 2010): 149–65.

Porter, Dorothy. *Health, Civilization and the State: A History of Public Health from Ancient to Modern Times*. New York: Routledge, 1999.

Prince, Ruth J. "Situating Health and the Public in Africa: Historical and Anthropological Perspectives." In Prince and Marsland, *Making and Unmaking Public Health*, 1–51.

Prince, Ruth J., and Rebecca Marsland, eds. *Making and Unmaking Public Health in Africa: Ethnographic and Historical Perspectives*. Athens: Ohio University Press, 2014.

Prins, Gwyn. "But What Was the Disease? The Present State of Health and Healing in African Studies." *Past and Present* 124 (1989): 159–79.

"Problems of African Native Diet." Special issue, *Africa* (Journal of the International Institute of African Languages and Culture) 9, no. 2 (April 1936).

Reid, Richard. *Political Power in Pre-colonial Buganda: Economy, Society and Warfare in the Nineteenth Century*. Athens: Ohio University Press, 2002.

Rice, Andrew. "The Peanut Solution: Could a Peanut Paste Called Plumpy'nut End Malnutrition?" *New York Times*, September 2, 2010, sec. Magazine, http://www.nytimes.com/2010/09/05/magazine/05Plumpy-t.html.

Richards, Audrey I. *The Changing Structure* of *a Ganda Village: Kisozi, 1892–1952*. Kampala, Uga.: East African Institute of Social Research, 1956.

———, ed. *Economic Development and Tribal Change: A Study of Immigrant Labour in Buganda*. Cambridge: W. Heffer & Sons Ltd., 1952.

———. *Hunger and Work in a Savage Tribe: A Functional Study of Nutrition among the Southern Bantu*. London: Routledge, 1932.

———. *Land, Labour and Diet in Northern Rhodesia: An Economic Study of the Bemba Tribe*. 1939; reprint, Hamburg: International African Institute Hamburg, 1995.

Richards, Audrey I., Ford Sturrock, and Jean M. Fortt, eds. *Subsistence to Commercial Farming in Present-Day Buganda: An Economic and Anthropological Survey*. Cambridge: Cambridge University Press, 1973.

Roberts, A. D. "The Sub-Imperialism of the Baganda." *Journal of African History* 3 (1962): 435–50.

Roscoe, John. *The Baganda: An Account of Their Native Customs and Beliefs*. London: MacMillan and Co., 1911.

Rosen, George. *A History of Public Health*. Extended ed. Baltimore: Johns Hopkins University Press, 1993.

Sachs, Jeffrey. "Saying 'Nuts' to Hunger." *Huffington Post*, September 6, 2010, http://www.huffingtonpost.com/jeffrey-sachs/saying-nuts-to-hunger_b_706798.html.

Schneider, William H. *The History of Blood Transfusion in Sub-Saharan Africa*. Athens: Ohio University Press, 2013.

Schoenbrun, David L. "Gendered Histories between the Great Lakes: Varieties and Limits." *International Journal of African Historical Studies* 29, no. 3 (1997): 461–92.

———. *A Green Place, A Good Place: Agrarian Change, Gender, and Social Identity in the Great Lakes Region to the 15th Century*. Portsmouth: Heinemann, 1998.

Schofield, Hugh. "Legal Fight over Hunger Wonder-Product." *BBC News*, April 8, 2010, sec. Europe. http://news.bbc.co.uk/2/hi/europe/8610427.stm.

Scott, Joan W. "The Evidence of Experience." *Critical Inquiry* 17 (1991): 773–97.

Shapiro, Karen. "Doctors or Medical Aids –The Debate over the Training of Black Medical Personnel: South Africa in the 1920s and 1930s." *Journal of Southern African Studies* 13, no. 2 (1987): 234–55.

Sicherman, Carol. *Becoming an African University: Makerere 1922–2000*. Trenton, NJ: Africa World Press, Inc., 2005.

Silla, Eric. *People Are Not the Same: Leprosy and Identity in Twentieth-Century Mali*. Portsmouth: Heinemann, 1998.

Snoxall, R. A., ed. *Luganda-English Dictionary*. Oxford: Clarendon Press, 1967.

Southall, Aidan W., and Peter C. W. Gutkind. *Townsmen in the Making: Kampala and Its Suburbs*. Kampala, Uga.: East African Institute of Social Research, 1956.

Southwold, M. "Ganda Conceptions of Health and Disease." In *Attitudes to Health and Disease among some East African Tribes*, 44–47. Kampala, Uga.: East African Institute of Social Research, 1959.

Stepan, Nancy Leys. *Picturing Tropical Nature*. Ithaca: Cornell University Press, 2001.

Stephens, Rhiannon. *A History of African Motherhood: The Case of Uganda, 700–1900*. Cambridge: Cambridge University Press, 2013.

Summers, Carol. "Grandfathers, Grandsons, Morality, and Radical Politics in Late Colonial Buganda." *International Journal of African Historical Studies* 38, no. 3 (2005): 427–47.

———. "Intimate Colonialism: The Imperial Production of Reproduction in Uganda, 1907–1925." *Signs: Journal of Women in Culture and Society* 16, no. 4 (1991): 792–93.

———. "Radical Rudeness: Ugandan Social Critiques in the 1940s." *Journal of Social History* 39, no. 3 (2006): 741–70.

———. "'Subterranean Evil' and 'Tumultuous Riot' in Buganda: Authority and Alienation at King's College, Budo, 1942." *Journal of African History* 47, no. 1 (2006): 93–113.

———. "Young Buganda and Old Boys: Youth, Generational Transition, and Ideas of Leadership in Buganda, 1920–1949." *Africa Today* 51, no. 3 (2005): 120–21.

Susser, Mervyn. "Pioneering Community-Oriented Primary Care." *Bulletin of the World Health Organization* 77, no. 5 (1991): 436–38.

Susser, Mervyn, and Zena Stein. *Eras in Epidemiology: The Evolution of Ideas.* Oxford: Oxford University Press, 2009.

Swanson, Maynard W. "The Sanitation Syndrome: Bubonic Plague and Urban Native Policy in the Cape Colony, 1900–1909." *Journal of African History* 18, no. 3 (1977): 387–410.

Tappan, Jennifer. "Blood Work and Rumors of Blood: Nutritional Research and Insurrection in Buganda, 1938–1952." In "Incorporating Medical Research into the History of Medicine in East Africa," special edition of *International Journal of African Historical Studies* 47, no. 3 (2014): 473–94.

Terris, Milton. "Concepts of Health Promotion: Dualities in Public Health Theory." *Journal of Public Health Policy* 13, no. 3 (1992): 267–76.

Thomas, Lynn. *Politics of the Womb: Women, Reproduction and the State in Kenya.* Berkeley: University of California Press, 2003.

Thompson, Gardner. "Colonialism in Crisis: The Uganda Disturbances of 1945." *African Affairs* 91 (1992): 605–24.

———. *Governing Uganda: British Colonial Rule and its Legacy.* Kampala, Uga.: Fountain Publishers, 2003.

Tilley, Helen. *Africa as a Living Laboratory: Empire, Development, and the Problem of Scientific Knowledge, 1870–1950.* Chicago: University of Chicago Press, 2011.

———. "Ecologies of Complexity: Tropical Environments, African Trypanosomiasis, and the Science of Disease Control in British Colonial Africa, 1900–1940." *Osiris* 19 (2004): 21–38.

Tripp, Aili Mari. *Women & Politics in Uganda.* Madison: University of Wisconsin Press, 2000.

Tripp, Aili Mari, and Joy C. Kwesiga, eds. *The Women's Movement in Uganda: History, Challenges and Prospects.* Kampala, Uga.: Fountain, 2002.

Tuck, Michael William. "Venereal Disease, Sexuality and Society in Uganda." In *Sex, Sin and Suffering: Venereal Disease and European Society since 1870,* edited by R. Davidson and L. Hall, 191–203. New York: Routledge, 2001.

Turshen, Meredith. *The Politics of Public Health*. New Brunswick, NJ: Rutgers University Press, 1989.
Twaddle, Michael. "The *Bakungu* Chiefs of Buganda Under British Colonial Rule, 1900–1930." *Journal of African History* 10 (1969): 309–22.
———. *Kakungulu and the Creation of Uganda*. London: James Currey, 1993.
———. "The Muslim Revolution in Buganda." *African Affairs* 71, no. 282 (1972): 54–72.
Jan Vansina. *Oral Tradition as History*. Madison: University of Wisconsin Press, 1985.
———.*Oral Tradition: A Study in Historical Methodology*. London: Routledge & Paul, 1965.
Vaughan, Megan. *Curing Their Ills: Colonial Power and African Illness*. Stanford: Stanford University Press, 1991.
———. "Healing and Curing: Issues in the Social History and Anthropology of Medicine in Africa." *Society for the Social History of Medicine* 7, no. 2 (1994): 283–95.
———. "Reported Speech and Other Kinds of Testimony." In White, Miescher, and Cohen, *African Words, African Voices*, 53–77.
———. "Syphilis in Colonial East Africa: The Social Construction of an Epidemic." In *Epidemics and Ideas: Essays on the Historical Perception of Pestilence*, edited by T. Ranger and P. Slack, 269–302. New York: Cambridge University Press, 1992.
Webb Jr., James L. A. "The Art of Medicine: The Historical Epidemiology of Global Disease Challenges." *Lancet* 385 (January 24, 2015): 322–23.
———. "The First Large-Scale Use of Synthetic Insecticide for Malaria Control in Tropical Africa: Lessons from Liberia, 1945-62." In Giles-Vernick and Webb Jr., *Global Health in Africa*, 42–69.
———. "Historical Epidemiology and Infectious Disease Processes in Africa." *Journal of African History* 54, no. 1 (2013): 3–10.
———. *Humanity's Burden: A Global History of Malaria*. New York: Cambridge University Press, 2009.
———. *The Long Struggle against Malaria in Tropical Africa*. New York: Cambridge University Press, 2014.
Webel, Mari. "Medical Auxiliaries and the Negotiation of Public Health in Colonial Northwestern Tanzania." *Journal of African History* 54, no. 3 (2013): 393–16.
Wendland, Claire L. *A Heart for the Work: Journeys through an African Medical School*. Chicago: University of Chicago Press, 2010.
———. "Research, Therapy, and Bioethical Hegemony: The Controversy over Perinatal HIV Research in Africa." *African Studies Review* 51, no. 3 (2008): 1–23.
West, Henry W. *Land Policy in Buganda*. Cambridge: Cambridge University Press, 1972.
White, Luise. *The Comforts of Home: Prostitution in Colonial Nairobi*. Chicago: University of Chicago Press, 1990.

———. *Speaking with Vampires: Rumor and History in Colonial Africa*. Berkeley: University of California Press, 2000.

———. "'They Could Make Their Victims Dull': Genders and Genres, Fantasies and Cures in Colonial Southern Uganda." *American Historical Review* 100, no. 5 (1995): 1379–402.

———. "True Stories: Narrative, Event, History and Blood in the Lake Victoria Basin." In White, Miescher, and Cohen, *African Words, African Voices*, 281–304.

White, Luise, Stephan F. Miescher, and David William Cohen, eds. 2001. *African Words, African Voices: Critical Practices in Oral History*. Bloomington: Indiana University Press, 2001.

Whyte, Susan Reynolds. "Medicines and Self-Help: The Privatization of Health Care in Eastern Uganda." In Hansen and Twaddle, *Changing Uganda*, 130–48.

———. *Questioning Misfortune: The Pragmatics of Uncertainty in Eastern Uganda*. Cambridge: Cambridge University Press, 1997.

———. ed., *Second Chances: Surviving AIDS in Uganda*. Durham, NC: Duke University Press, 2014.

Whyte, Susan Reynolds, Michael A. Whyte, Lotte Meinert, and Jenipher Twebaze. "Therapeutic Clientship: Belonging in Uganda's Projectified Lanscape of AIDS Care." In Biehl and Petryna, *When People Come First*, 140–65.

Widdowson, Elsie. M. "Obituary Notice: R. A. McCance (9 December 1898–5 March 1993)." *Proceedings of the Nutrition Society* 52 (1993): 383–86.

Williams, A. W. "The History of Mulago Hospital and the Makerere College Medical School." *East African Medical Journal* 29 (July 1952): 253–63.

Worboys, Michael. "The Comparative History of Sleeping Sickness in East and Central Africa, 1900–1914." *History of Science* 32 (1994): 89–101.

———. "The Discovery of Colonial Malnutrition Between the Wars." In *Imperial Medicine and Indigenous Societies*, edited by David Arnold, 208–25. New York: Manchester University Press, 1988.

———. "Tropical Diseases." In *Companion Encyclopedia in the History of Medicine*, edited by William F. Bynum and Roy Porter, 512–13. London: Taylor & Francis, 1993.

World Health Organization. "Health Expenditure Per Capita (Current US$)." World Bank Group. Accessed August 23, 2014. http://data.worldbank.org/indicator/SH.XPD.PCAP/countries?display=default.

Wrigley, Christopher. "Bananas in Buganda." *Azania* 24 (1989): 60–80.

———. "The Changing Economic Structure of Buganda." In Fallers, *The King's Men*, 16–63.

Wright, Marcia. *Strategies of Slaves and Women: Life-Stories from East/Central Africa*. New York: Lilian Barber Press, 1993.

Wylie, Diana. "The Changing Face of Hunger in Southern African History, 1880–1980." *Past and Present* 22 (1989): 159–99.

———. *Starving on a Full Stomach: Hunger and the Triumph of Cultural Racism in Modern South Africa*. Charlottesville: University of Virginia Press, 2001.

Interviews

Babirye, Cate. Interviewed by Jennifer Tappan and Hajjarah Nambwayo (translator). Recording. Bamunanika, Uganda, July 19, 2012.

Baptist, John. Interviewed by Jennifer Tappan and Jemba Enock Kalema (translator). Recording. Luteete, Uganda, July 23, 2004.

Church, Mike, and Paget Stanfield. Interviewed by Jennifer Tappan. Recording. Edinburgh, Scotland, November 26, 2003.

Daamba, Namakula. Interviewed by Jennifer Tappan and Hajjarah Nambwayo (translator). Recording. Luteete, Uganda, July 18, 2012.

Haswell, Margaret Rosary. Interviewed by Jennifer Tappan. Recording. Oxford, England, November 20, 2003.

Kakitahi, John. Interviewed by Jennifer Tappan. Recording. Makerere Medical School, Mulago Hospital, Kampala, June 23, 2004.

Kaloli, Nabanja Federesi. Interviewed by Jennifer Tappan and Ssennoga Jackson (translator). Recording. Luteete, Uganda, May 6, 2004.

———. Interviewed by Jennifer Tappan and Ssennoga Jackson (translator). Recording. Luteete, Uganda, June 17, 2004.

———. Interviewed by Jennifer Tappan and Hajjarah Nambwayo (translator). Recording. Luteete, Uganda, July 25, 2012.

Kayemba, Robinah Namulindwa. Interviewed by Jennifer Tappan and Jemba Enock Kalema (translator). Recording. Luteete, Uganda, June 23, 2004.

Kazibwe, Keziya. Interviewed by Jennifer Tappan and Ssennoga Jackson (translator). Recording. Luteete, Uganda, June 16, 2004.

Kiwanuka, Batister. Interviewed by Jennifer Tappan and Hajjarah Nambwayo (translator). Recording. Luteete, Uganda, July 18, 2012.

Kwene, Mark. Interviewed by Jennifer Tappan. Recording. Mwanamugimu Nutrition Unit, Kampala Uganda, February 2004.

Kyambadde, Dorobina, and Alex Kimbowa. Interviewed by Jennifer Tappan and Hajjarah Nambwayo (translator). Recording. Luteete, Uganda, July 20, 2012.

Kyaze, Florence Joyce, and Wilson Ssalongo Kyaze. Interviewed by Jennifer Tappan and Ssennoga Jackson (translator). Recording. Luteete, Uganda, April 2004.

———. Interviewed by Jennifer Tappan and Ssennoga Jackson (translator). Recording. Luteete, Uganda, June 2, 2004.

Kyeyunne, Bumbakali. Interviewed by Jennifer Tappan and Ssennoga Jackson (translator). Recording. Butto, Uganda, July 21, 2004.

Kyeyunne, Kasifa, and Yahaya Kyeyunne. Interviewed by Jennifer Tappan and Ssennoga Jackson (as translator). Recording. Butto, Uganda, May 21, 2004.

Mulindwa, "Ssalongo" Stephen Maseruka. Interviewed by Jennifer Tappan. Recording. Luteete, Uganda, July 17, 2012.

Musoke, Ephraim. Interviewed by Jennifer Tappan and Ssennoga Jackson (translator). Recording. Luteete, Uganda, April 2004.

———. Interviewed by Jennifer Tappan and Ssennoga Jackson (translator). Recording. Luteete, Uganda, May 13, 2004.

———. Interviewed by Jennifer Tappan. Recording. Luteete, Uganda, June 3, 2004.

Musoke, Ephraim, and Catherine Musoke. Interviewed by Jennifer Tappan and Hajjarah Nambwayo (translator). Recording. Luteete, Uganda, July 23, 2012.

Musoke, Philipa. Interviewed by Jennifer Tappan. Recording. Makerere Medical School, Mulago Hospital, Kampala, Uganda, January 5, 2004.

Muwazi, Louis Mugambe. Interviewed by Jennifer Tappan. Recording. Makerere Medical School, Mulago Hospital, Kampala, Uganda, March 2004.

Nadamba, Kasalina. Interviewed by Jennifer Tappan and Ssennoga Jackson (translator). Recording. Luteete, Uganda, May 26, 2004.

Nakasiko, Christine. Interviewed by Jennifer Tappan and Hajjarah Nambwayo (translator). Recording. Kisanku, Uganda, July 21, 2012.

Nakeebe, Agnes, Florence Nakalyowa, and Milly Nalubega. Interviewed by Jennifer Tappan and Hajjarah Nambwayo (translator). Recording. Luteete, Uganda, July 18, 2012.

Nakidalu, Joyce Lukwago. Interviewed by Jennifer Tappan and Hajjarah Nambwayo (translator). Recording. Kisanku, Uganda, July 21, 2012.

Nakyejwe, Daisy. Interviewed by Jennifer Tappan and Jemba Enock Kalema (translator). Recording. Luteete, Uganda, July 23, 2004.

———. Interviewed by Jennifer Tappan and Hajjarah Nambwayo (translator). Recording. Luteete, Uganda, July 20, 2012.

Nalwoga, Harriet. Interviewed by Jennifer Tappan and Hajjarah Nambwayo (translator). Recording. Luteete, Uganda, July 18, 2012.

Namakula, Madina. Interviewed by Jennifer Tappan and Hajjarah Nambwayo (translator). Recording. Kisanku, Uganda, July 24, 2012.

Namakula, Nula, and Robinah Nanono. Interviewed by Jennifer Tappan and Hajjarah Nambwayo (translator). Recording. Kisanku, Uganda, July 24, 2012.

Namboze, Josephine. Interviewed by Jennifer Tappan. Recording. Kampala, Uganda, July 12, 2004.

Namusoke, Budesiaa. Interviewed by Jennifer Tappan and Jemba Enock Kalema (translator). Recording. Luteete, Uganda, July 24, 2004.

Namusoke, Emerisian. Interviewed by Jennifer Tappan and Hajjarah Nambwayo (translator). Recording. Luteete, Uganda, July 18, 2012.

Namusoke, Maria Zerenah. Interviewed by Jennifer Tappan and Hajjarah Nambwayo (translator). Recording. Kisanku, Uganda, July 24, 2012.

Nankabirwa, Fatuma. Interviewed by Jennifer Tappan and Hajjarah Nambwayo (translator). Recording. Bamunanika, Uganda, July 19, 2012.

Nansamba, Catherine. Interviewed by Jennifer Tappan and Hajjarah Nambwayo (translator). Recording. Luteete, Uganda, July 20, 2012.

Nanteza, Robinah. Interviewed by Jennifer Tappan and Jemba Enock Kalema (translator). Recording. Luteete, Uganda, July 2004.

Nanyonga, Solom. Interviewed by Jennifer Tappan and Hajjarah Nambwayo (translator). Recording. Luteete, Uganda, July 19, 2012.

Nanziri, Peragiya. Interviewed by Jennifer Tappan and Hajjarah Nambwayo (translator). Recording. Bamunanika, Uganda, July 19, 2012.
Ndugwa, Chris. Interviewed by Jennifer Tappan. Notes. Makerere Medical School, Mulago, Kampala, Uganda, June 8, 2004.
Njovu, Patrick. Interviewed by Jennifer Tappan. New York, New York, January 9, 2007.
Odonga, Alexander. Interviewed by Jennifer Tappan. Recording. Kampala, Uganda, 2004.
Pascal, Mposa. Interviewed by Jennifer Tappan and Jemba Enock Kalema (translator). Recording. Luteete, Uganda, July 24, 2004.
Sempa, Christopher. Interviewed by Jennifer Tappan and Jemba Enock Kalema (translator). Recording. Luteete, Uganda, July 24, 2004.
Ssemujju. Interviewed by Jennifer Tappan and Ssennoga Jackson (translator). Recording. Luteete, Uganda, May 26, 2004.
Stanfield, Paget. Interviewed by Jennifer Tappan. Recording. Edinburgh, Scotland, November 27, 2003.
Walujjo, Albert. Interviewed by Jennifer Tappan and Jemba Enock Kalema (translator). Recording. Luteete, Uganda, June 2004.
Wamala, Julie. Interviewed by Jennifer Tappan. Mwanamugimu Unit, Mulago Hospital, Kampala Uganda, July 30, 2012.
Whitehead, Roger. Interviewed by Jennifer Tappan. Recording. Kamapala, Uganda, December 9, 2003.

INDEX

60 Minutes, 139

accumulated reflections, 6, 35
acquired immunodeficiency syndrome (AIDS), xi, 130–32; Global Fund, 131; policy, 133–34; project, 127; research, 130–33; The AIDS Support Organization (TASO), 131; treatment. *See* United States President's Emergency Plan for AIDS Relief (PEPFAR)
activism, 23, 66, 141
Adichie, Chimamanda Ngozi, 5
aflatoxin, 24, 50, 51
Africanization, 80, 114
agriculture, 8, 63, 93, 102, 116. *See also* cash crops; consumption crops, 122
AIDS. *See* acquired immunodeficincy syndrome (AIDS)
Ainsworth, Mary, 54–61
akaboxi, 85, 91
Algeria, 50
Amin, Idi, 63, 102, 111, 113, 118–28, 130, 135
anemia, 16, 18, 40; testing, 17
anorexia, 2, 41, 45
antibiotics, 69, 141
apartheid, 71
appetite, 32, 33
Archives of Disease in Childhood, 15
Arlac, 50
Association of Physicians of East Africa, 115
autocracy, 24, 126
autopsy, 16, 21, 22, 23

The Baby Food Tragedy, 65
The Baby Killer, 65
balance bed, 28–30
Bamunanika palace, 99, 117
banana. *See* matooke (banana or plantain); leaf. *See* luwombo (banana leaf)
basawo (healers), 30, 128
Belgian Congo, 67
Biafra War, 5
bibanja (garden), 85, 102, 106
biochemist, 18, 19, 28, 113
biomedicine, 5–6, 25, 34–35, 42, 57, 60, 64, 68, 74, 80, 80–82, 87, 92, 142; training, 7, 36
biopsy, 16, 18–19, 23, 26
bismuth, 31
bloodletting, 28, 29–30, 35
blood tests, 16, 18–19, 23, 25, 26, 27–28, 31, 34
Bombay, 43
Botswana, 129
bottle feeding, 54–67, 68, 141
Brazil, 22, 50
breastfeeding, 14, 33, 44, 45, 46, 51, 54–67; weaning, 45, 46, 50, 53, 59, 68, 81, 106, 109
Briend, André, 137, 138
British Protectorate, the, 8, 12, 25, 76; High Commission, 118
broad-spectrum approach, 114
Buganda, 7–10, 30, 44, 45, 46, 57, 72, 81, 99, 102, 117, 122, 124
Burundi, 17, 43, 63
Bush War, 104, 107, 123, 126, 130

Cairo, 16
calendars, 88–89
cash crops, 8, 32, 49, 63, 102, 122. *See also* agriculture; cotton, 8, 63
Casilin, 38
cassava, 46, 47
central Ugandan Hospital. *See* Mulago
Chad, 138
child development. *See* growth rates
child welfare clinics; Child Welfare Clinic (MRC), 27; maternal and, 31, 99; mobile (Welbourne's), 53, 57, 58, 71, 82
chloroquine, 69
Church, Mike, 41–42, 74, 76, 78, 81, 83, 91, 92–93, 96, 99–100
Church Missionary Society (CMS), 8, 12–13, 16, 99
Church World Service, 53
Cinva Ram (brickmaking device), 100, 104
CMS. *See* Church Missionary Society (CMS)

colonialism, 5, 7, 8, 11, 15, 23, 24, 30, 34–35, 72; medicine, 35–36; post, xiii, 24, 68, 111, 113, 117
commerce, 8, 50, 51, 116; expansion, 63; marketing, 65
Community Health and AIDS Project, 127
Congo-Brazzaville, 129
consultants, 49
consumption crops, 122
contamination, 65, 137
Cook, Albert, 8, 12, 16, 99; Medical Library, xvii
Cook, Catherine, 99
Cooper, Anderson, 139, 141
cotton, 8. *See* cash crops, cotton
cough, 55–56
Crane, Johanna, 132

Davies, Jack, 20–22, 43–44, 61
Dean, Rex, 26, 27, 28, 29, 31, 35, 37, 38, 39, 42, 43, 44, 46, 49–50, 51, 53, 66, 68–69, 81, 113, 137, 141
decentralization, 112, 128
dehydration, 40, 121
demography, 9, 12, 116, 125
dermatosis, 2, 41
deworming, 31
diagnostics, 11–36, 42, 71
diarrhea, 33, 38, 55, 56, 57, 61–62, 65
dictatorship, 112, 122, 124–25, 130
dietary discipline, 39–40
diphtheria, 71
Diseases of Children in the Subtropics and Tropics, 62
drug; antiretroviral, 131; chloroquine, 69; commercial provisions, 136; companies. *See* pharmaceutical companies; kits, 121; provisions, 127, 128
DSM. *See* milk formulas, dried skim (DSM)
Dunn Nutrition Unit in Cambridge, 120

Ebine, George, 118
economics, 4, 24, 49, 82, 100–102, 125, 129, 130; cash-crop, 8; development, 62, 63; home, 116; inflation, 126; middle class, 63; reform, 112; war, 122
edema, 1, 20, 27, 32
Edesia, 139
education. *See* public health, education
ekigalanga, 33
Emiru, Vicent, 118
endwadde ez'ekiganda (Ganda diseases), 30
endwadde ez'ekizungu (European diseases), 30
England, xii, 8, 16, 20, 76, 118
Entebbe airport, 120
enva (sauces), 45, 46, 84
enzyme, 26, 32; synthesis, 20
epidemic, 9, 112, 130, 131, 135. *See also* acquired immunodeficiency syndrome (AIDS); *See also* human immunodeficiency virus (HIV)
epidemiology, 62; historical, 6, 10; malnutrition, 4
epistemology, 80
ethics, 27, 31, 34, 35, 140
Ethiopia, 138
etiology, 11, 18, 20, 21, 42, 114
ettu paste, 82–84
Expert Committee on Nutrition (WHO/FAO), 21, 43, 50, 66

faddism, 112, 135
family planning, 71
famine, 136, 138, 143
FAO. *See* United Nations (UN), Food and Agriculture Organization (FAO)
feeding tube, 22, 32, 70; intragastric, 41
fever, 33, 55, 57
food insecurity, 133
food preparation, 75–84, 106–10, 122, 137; demonstration, 77, 78, 93–94, 102, 106–8
foreign aid, 126, 129, 131
Fortifex, 50

Gambia, 28
garden. *See* kibanja (garden)
gastroenteritis, xiv, 54, 61
Gates Foundation, 131
Gayaza High School, 78, 82
Geneva, 21
Georgia, 139
Germany, 38, 82
Ghana, 14
Gilman, Theodore and Joseph, 19
Global Fund to Fight AIDS, Tuberculosis and Malaria, 131
global health. *See* public health, global
Global Health in Africa, 7
glucose, 57
Golden, Michael, 137

Graboyes, Melissa, 6
Great Depression, 13, 99
Great Ormond Street Sick Children's Hospital, 73
Great Protein Fiasco, The, 64–65
groundnut biscuit, 50–51
growth rates, 18, 40–41, 44, 47, 57, 92, 93; charts, 93–94
Guatemala's Institute of Nutrition for Central America and Panama (INCAP), 49, 50

Haiti, 141
healers (basawo), 30, 128
health expenditure, per-capita, 69
Health, Hunger and Humanity grant (Rotary International), 123, 130
height, 18, 92
heritage. *See* local practices
hierarchy, 24–25
high-protein. *See also* therapy, high-protein; diet, 14; foods, 45, 49–50, 53, 64–65, 66–67, 68, 82–83; formula, 37–41, 43, 48–49; mixtures, 50, 82, 85
high-protein food program, 53, 64–66, 141
Himsworth, Harold, 19–20, 22
historians, xiii, 12, 24, 35, 35–36, 128; colonial medicine, 5; epidemiological, 6, 10; global health, 5; malnutrition, 11
hospitals, 127–28, 128; Great Ormond Street Sick Children's, 73; Mengo. *See* Mengo Hospital; Mulago. *See* Mulago Hospital
How to Feed Your Child, 82
human immunodeficiency virus (HIV), xi, 3, 111–34, 141; policy, 133–34; research, 127, 132; treatment, 133
humanitarian aid. *See* philanthropy
hunger-wonder product. *See* Plumpy'Nut
hygiene, 98, 114

Iliffe, John, 24, 128, 129
IMF. *See* International Monetary Fund (IMF)
immigrants, 17–18, 43, 63
immunizations, 6, 71, 127, 131, 141; trial, 131–32
INCAP. *See* Guatemala's Institute of Nutrition for Central America and Panama (INCAP)
Incaparina, 50
independence. *See* sovereignty
India, 50
Infectious Diseases Institute (IDI), 132
influencers, 96–110
insurrection. *See* political unrest
intergenerational knowledge transfer, xiv
International Monetary Fund (IMF), 125, 126, 130; conditionalities, 125; Economic Recovery Package, 126
interview, xi, xiii, xiv, 53, 72, 85, 100, 104, 106–8. *See also* testimony
Israel, 120

Jelliffe, Derrick B., 53, 60–61, 63, 65, 72–73, 81, 82, 83, 89
Jézéquel, Jean-Hervé, 141

kabaka (king), 8, 22, 24, 81, 99, 117
kadongo kamu (ballad singer), 89
Kafu River, 8
Kagera River, 7
Kakitahi, John, 87, 119, 120, 121, 122, 123, 125, 129–30, 130, 135
Kaleeba, Noerine, 131
Kaloli, Nabanja, xi, 98, 99, 103, 107, 109, 134
Kampala, xi, 7, 12, 18, 53, 55, 57, 60, 63, 71, 99, 109, 122, 130, 131, 135
Karamoja, 136
Kasangati Health Center, 71, 73, 100, 122, 123
Kasozi, A.B.K., 125
Kayemba, Robinah, 106
Kayunga, 122, 123
Kenya, 8, 9, 13, 31, 60, 119; Northern Frontier District, 15
kibanja (garden), 81, 93, 128
Kiboneka, Elizabeth, 132, 133
Kibukamusoke, John, 115–16, 118
king. *See* kabaka (king)
kitobero, 68–110, 106, 122, 132, 134, 135, 141, 142; plays, 89, 92. *See also* A Play about Kitobero; *See also* The Story of a Builder; song, 83–84, 89, 91
kwashiorkor, 3, 4, 14, 20, 33, 40, 42, 98. *See also* therapy, high-protein; change in treatment of, 74; complicated, 70; description of, 1; emergency action campaign, 48; etiology of, 21; international instances of, 22; reoccurance of, 69; skim-milk based therapy for, 62; therapy for, 44, 49, 51; undernutrition versus, 64

Kwashiorkor in Africa, 21, 44, 48
Kyalwazi, Sebastian, 119
Kyambadde, Dorobina, 134
Kyeyunne, Bumbakali, 103
Kyeyunne, Kasifa, 106
Kyobe, John, 16

labor strike, 23
Lac-Tone, 50
lactose intolerance, 38
Lancet, 14, 40, 64
Latham, Michael, 60
League of Nations (LON), 13
lecture-poster barrier, 87–95
Lehmann, Ferdie, 18
Lescanne, Michel, 136, 137
Lives in Peril, 48
living memory, 106
local anxieties, 31–32, 34, 54, 57, 71, 87, 123; economic, 125
local practices, 45, 72, 73, 81, 106, 107, 108, 109, 110, 135, 143; cooking, 82–83; diet, 44–45, 45, 46, 46–47, 49, 81, 141–42; feeding, 78; healing, 30; spiritual, 33; weaning. *See* breastfeeding, weaning
LON (League of Nations), 133
Luganda, 42, 82, 84
lukiiko (council), 100, 102
Lukwago, Faith, 89, 120
lumonde (sweet potato), 46, 82, 84
Luo, 102
Luteete Health Center, xi–xiv, 13, 53, 85, 98, 99, 100, 102, 103, 104, 106, 107, 108, 109, 110, 111, 112, 113, 117, 118, 121, 122, 123, 124, 135
Luwero Health Center, 124
Luwero Triangle, 124, 125
luwombo (banana leaf), 82, 83, 85

magendo (smuggling), 130
magezi (wisdom), 92
magic-bullet, 43, 64, 82, 136, 141, 142
maize, 140
Makerere Medical School, 12, 15, 16, 25, 72, 73, 96, 114, 116, 127, 129, 132; Department of Agriculture, 102; Department of Preventive Medicine, 71, 72; Graduates Association, 24; Institute of Public Health, 119
malaria, 4, 40, 131; medication, 121
Malawi, 14, 140
Malaya, 51
Malaysia, 15
malnutrition, 27, 33, 45, 54, 61, 68, 71, 74, 94, 99, 102, 106, 114, 133, 134, 135, 136–43; acute childhood, xi, xii, 1, 17, 64, 100, 130; commerciogenic, 65; early history of, 11; ecology of, 72; environmental, 47, 71; global problem, 1, 64, 139; medicalization of, 34, 37–67, 68–69, 70–71, 78, 109, 136, 141, 142; protein-calorie (PCM), 64; protein-energy (PEM), 64; relapse, 53–54, 68–80; severe acute (SAM), 1, 2, 3, 5, 10, 11, 12, 13, 20, 32, 34, 35, 37–38, 43, 46, 47, 48, 66, 68–69, 70, 80, 94, 102, 107, 109, 110, 119, 136, 137, 139, 142. *See also* kwashiorkor; signs of, 92
MANA (Mother Administered Nutritive Aid), 139
Manary, Mark, 140
Maradi, 138
marasmus (undernutrition), 1, 2, 3, 4, 54, 61, 64, 66, 133. *See also* malnutrition
Marxism, 126
Maseruka Mulindwa, Stephen, 108, 121, 124, 125, 135
maternal deprivation, 45
matooke (banana or plantain), 46, 81, 82
Mauritania, 137
McLaren, Donald S., 64
meal frequency, 45
measles, 71
Médecins Sans Frontières (MSF), 136, 138, 139, 141; therapeutic feeding centers, 138
media, 116
medical officer, xix, 23, 24, 74, 128; colonial, 14, 20, 25, 72
Medical Research Council (MRC), 22, 26, 27, 28, 29, 38, 44, 46, 49, 50, 113, 115–16, 118, 119–20; Child Nutrition Unit, 114, 115; Infantile Malnutrition Unit, 26, 39, 41, 43, 114
memories; living, xiv; local, xiii
Mengo Hospital, 31, 71, 99
metabolism, 19
micronutrients, 14
midwives, xii, 16, 99, 107, 117
Mildmay Uganda, 133
milk formulas, 67, 69, 75. *See also* protein, formulas; cost of, 60; cottonseed oil

and, 38, 49, 51; cream, 51; distribution, 53, 54, 57, 62, 66, 69, 141; dried skim (DSM), 38, 42, 48, 50, 51, 52, 53, 59, 63, 66, 67, 69, 82, 137, 140, 142; F-75, 137; F-100, 137; high-protein, 49; intolerance. *See* lactose intolerance; liquid, 137; preparation, 69, 73; protein concentrate, 38, 40. *See also* Casilin; reinforced packets and labels, 51, 52, 54; substitutes, 48; therapy, 49
Millennium Development Goals, 136
minerals, 14–15, 137, 138, 140
Ministry of Health, 131
Ministry of Social Development, 98
miruka (chief), 100, 107, 117
missionaries, 8, 139; medical, 12
Moore, Mark, 139
morbidity, 3, 133
morgue, 21
mortality, 3, 23, 25, 27, 37, 38, 40, 44, 65, 129, 133, 134, 136; child, 4; infant, 66, 125
Mother Administered Nutritive Aid (MANA), 139
Mother and Child, 96–98
mothers, 14
MRC. *See* Medical Research Council (MRC)
MSF (Médecins Sans Frontières). *See* Médecins Sans Frontières (MSF)
Muganda, 16
Mugula, Dan, 89–90
Mukasa, Florence, xii, 107, 109
Mukasa, Luka, 100, 102
Mulago, 51
Mulago Hospital, xii, xiv, 7–10, 12, 17–23, 25–28, 32, 35–36, 41, 43, 51, 53, 54, 60–63, 69, 70, 74, 79, 85, 87, 98, 100, 106–7, 111–13, 119–21, 127, 128, 130, 135, 136, 140; certificates, 89, 90; Government Dispensary, 73; Infectious Diseases Institute (IDI), 132; maternity center, 107; nutrition rehabilitation program, 73, 76–77, 78, 80, 89, 94, 107, 109, 110, 123, 129, 132–33, 134; staff, 79, 88, 125, 142
Muller, Mike, 65
Museveni, Yoweri, 123, 126, 131
Musoke, Ephraim, xi, 49, 61, 103, 111
Musoke, Latimer, 60, 120, 121, 122, 125, 135
Muwazi, Eria, 16, 17, 18, 22–25, 34–35, 117
Mwanamugimu Nutrition Rehabilitation Program (Mwanamugimu Unit), ix, xii, xiii, xiv, xviii, 75, 77, 79, 80, 84, 85, 87–92, 94, 96, 98–100, 102, 106–10, 116–23, 125, 128, 129, 130, 132–35, 140, 142, 143, 147, 167–69, 171, 172, 177, 183, 202, 208, 210

Nairobi, 13, 15, 18
nakati, 102, 103
Namboze, Josephine, 55, 60, 71–73, 127, 129
Namibia, 130
Namusoke, Maria Zerenah, 85, 134
Nankabirwa, Fatuma, 85–86
Nansamba, Catherine, 60, 87
Nanteza, Robinah, 106, 107
Nanyonga, Solom, 87
Nanziri, Peragiya, 85, 107, 134
National Institute of Health (NIH), 132
National Research Council, 47; Committee on Protein Malnutrition, 47
National Resistance Army (NRA), 123, 124
Ndugwa, Chris, 119
negative feedback loop, 133
neoliberalism, 112, 126, 135
Nestlé, 65
New Internationalist, 65
New York Times, 141
NGO. *See* non-governmental organizations (NGO)
ngozi, xii, 33, 80, 169
niacin, 16
Niger, 137, 138, 139, 140
Nigeria, 5, 50
NIH. *See* National Health Institute (NIH)
Nile River, 7
nitrogen balance studies, 19, 23, 28, 47
nkejji (cichlids), 85
non-governmental organizations (NGO), 112, 127, 127–29, 128, 128–29, 131, 135, 136, 137; Concern Worldwide, 138
Novofood, 136–37; distribution, 136, 137
NRA (National Resistance Army), 123, 124
Nutella, 137
Nutriset, 137, 138, 139; licensed subsidaries of, 140; patents, 139
nutrition, xi, 2, 7, 67, 82, 98, 130, 133, 134, 137, 141, 142, 143. *See also* research, nutrition; applied, 116; childhood, 82; disorders, 21, 69; education. *See* public health, education; history, 70;

nutrition (*cont'd*)
intervention, 133; mother-provided, 68–110, 138, 139, 141; rehabilitation program, xiii, 7, 10, 70, 73, 80, 98, 111, 112, 113, 114, 123, 128, 133. *See also* Mulago Hospital, Mwanamugimu nutrition rehabilitation program; *See also* Mwanamugimu nutrition rehabilitation program; rehabilitation unit, 100, 102; Ugandan school of, 116; weaning, 68, 84
nutrition-infection complex, 114, 133. *See also* Whitehead, Roger
nutritionists, 45, 46, 137, 141

Obote, Milton, 117, 118, 123–24, 124, 125, 126, 127, 130, 135
obulwadde, 30
obusulo, 33
obwosi, 33
Odonga, Alexander, 119
olbuwadde bw'eccupa (bottle disease), 42, 67
oligarchy, 8, 24
olumbe, 30–36, 41
olutabu, 84
omusana, 33
Operation Sahel, 137
ophthalmology, 118
orphans, 38, 134
Oxfam, 99–100, 124

PAG. *See* United Nations (UN), Protein Advisory Group (PAG)
pancreatic atrophy, 20, 21, 22
pathology, 16, 20, 21, 66
PCM. *See* malnutrition, protein-calorie (PCM)
Peanut Butter Debate, 138
pediatrics, 15, 18, 25, 42, 53, 60, 61, 70, 71, 72, 73, 89, 119, 120, 133
pellagra, 14, 15; controversy, 18; infantile, 15
PEM. *See* malnutrition, protein-energy (PEM)
penicillin, 40, 141
PEPFAR. *See* United States President's Emergency Plan for AIDS Relief (PEPFAR)
per-capita health expenditure, 140
pertussis, 71
Pfizer, 132
pharmaceutical companies, 122, 128, 131, 132; Pfizer, 132
philanthropy, 5, 137
physicians, xiii, 18, 25, 30, 32–33, 36, 45, 53, 57, 61, 63, 66, 67, 68, 69, 72, 76, 81, 102, 109, 117, 119, 121, 122, 127, 128, 141
physiology, 18, 19
plaintain. *See* matooke (banana or plantain)
A Play about Kitobero, 89, 91
Plumpy'Doz, 138–39. *See also* ready-to-use therapeutic food (RUTF); distribution, 139
Plumpy'Nut, 138, 138–41. *See also* ready-to-use therapeutic food (RUTF); packets, 142; Revolution, 138; sales, 139–40
policies, 6, 10, 123, 127; fiscal, 125; No Survey without Service, 25, 35; pronatalist, 67
polio, 71
political unrest, xiii, 11, 22–25, 26, 27, 30, 34, 99, 104, 110, 111–34, 129, 130, 135, 136, 142, 143
politicians, 24, 117, 118
post-mortem examination. *See* autopsy
potassium, 40
poultry, 102
poverty, 50, 53, 61, 100, 102, 125, 128, 129, 131, 133, 140; reduction of, 126
practitioners; biomedical, 4, 42, 46, 54, 62, 64, 69, 78, 109; global health, 35, 71; private, 57
pregnancy, 12, 33, 46, 55, 93, 130
prevention, 25, 69, 71, 73, 94, 102, 122, 128, 131, 136, 138, 139, 140, 141, 142, 143; education, 143; long-term, 143; methodology, 80; practical, 78
privatization, 112
Project Peanut Butter in Malawi, 140. *See also* Manary, Mark
Pronutro, 50–51
protein, 20, 102; deficiency, 14, 20, 22, 27, 42, 43, 44, 45, 46, 47, 48, 50, 54, 66, 82. *See also* kwashiorkor; *See also* malnutrition, protein-calorie (PCM); *See also* malnutrition, protein-energy (PEM); evaluations, 17, 18, 27, 28; food program, 141; formulas, 37, 40, 41, 43. *See also* milk formulas; non-milk based therapy, 49–50; research. *See* research, protein; therapy, 38, 39, 43, 50; vegetable, 38,

51, 82, 83; worldwide gap, 47–53, 64, 66, 68
protest, 129
psychology, 44, 54, 116
public health, 6, 13, 44, 66, 80, 81, 110, 113, 125, 135, 142; campaigns, 45, 48, 51–53, 80, 82, 89, 106, 138; community education, 88; comprehensive, 68, 99–100, 136; education, 69, 72, 73, 81, 83, 87, 92, 116, 121, 126, 143; experts, 6; faddism, 112, 135; funding, 142; global, 35, 36, 68, 131, 132, 133, 135, 136, 139, 143; historians, 5; maternal education, 69–110, 121–22, 134, 138, 142; models of, 73; practical, 78; programs, 5, 7, 135; workers, 121

rash, 41
ready-to-use therapeutic food (RUTF), 137–43; distribution, 138, 139, 140, 141; locally produced, 140; manufacturing of, 139, 140; trials, 138
Red Cross, 124; French, 137; Trucks of Hope, 137
refugees, 124, 143; camps, 5
research, 6, 116; biomedical, 11, 36, 109; colonial medical, 35; history, 36, 71, 112–13; nutrition, 6, 7, 11, 22, 23, 25, 26, 27, 30, 34, 35, 36, 43, 70, 113, 119, 132; protein, 43; vitamin, 43
RESTORE Uganda, 133
Reynolds Whyte, Susan, 129
Rockefeller Foundation, 49, 102
Rotary International, 123, 130
rural outreach, 12, 13, 31, 63, 71, 96–110, 108, 110, 114, 116, 120, 123, 129–30, 130, 131, 142
RUTF (ready-to-use therapeutic food). *See* ready-to-use therapeutic food (RUTF)
Rwanda, 17, 43, 63, 137

Salem, Navyn, 139
SAM. *See* malnutrition, severe acute (SAM)
SAP. *See* structural adjustment program (SAP)
Save the Children Fund (SCF), 124; Nutrition Rehabilitation Unit, 73
SCF. *See* Save the Children Fund (SCF)
Schneideman, Ian, 72, 78
scientists, xiii, 6, 18, 32–33, 36, 45, 63, 66, 67, 68, 116; nutritional, 37
Sembeguya, George, 118
sociology, 116
South Africa, 50–51, 71, 72, 73
South Carolina, 139
sovereignty, 113–19, 135, 142, 143
soy, 49; blended, 140
Stanfield, Paget, 72, 74, 76, 78–79, 81, 92–93, 99–100, 113–14, 116, 118, 142
Stannus, Hugh, 14, 15
starvation, 65
Stokes, Gladys, 78, 80, 81, 82, 83, 89, 100, 120, 135, 142
The Story of the Builder, 92
structural adjustment program (SAP), 125, 126
Sudan, 69
Superamine, 50
supplements, 47, 54–55, 57, 58–59, 61, 139
Surpro, 50
Swahili, 42, 102
Switzerland, 65
syphilis, 12, 13, 16, 30–31

tabula, 84
Tanzania, 10, 60, 122, 130
TASO (The AIDS Support Organization), 131
technology, 10, 47, 121, 137
territorial expansion, 8
testimony, xiii, xviii, 54, 55, 57, 85, 106, 134. *See also* interview
therapy, 11, 42, 141; antiretroviral, 131, 133, 134, 141. *See also* acquired immunodeficiency syndrome (AIDS); drug, 39; high-protein, 38–41, 42–43, 48, 50, 66–67; holistic, 71, 72, 78, 116; home, 70; hospital, 30, 33, 42–43, 70, 74; kwashiorkor, 62; nutrition, 137, 138; ready-to-use therapeutic food (RUTF), 141; rehabilitation, 68–80, 74. *See also* nutrition, rehabilitation program
Time Life Magazine, 5
trauma, psychological, 45
treatments for, 137
Triassic Palace. *See* Mulago Hospital
Tripp, Aili, 98
Trowell, Hugh, 13, 14, 15, 16, 17, 18, 19, 20, 21, 22, 25, 26, 33, 43, 44, 62
Trucks of Hope, 137
tube-feeding, 32–33; intragastric, 41, 42; intravenous, 42

tuberculosis, 131
Tusitukirewamu club, 102, 103

Uganda; AIDS Commission, 131; Co-Operative Creameries, 51; Food and Nutrition Council, 115; government, 24, 114, 127–28, 129; Health Policy Review Commission, 127; Medical Association, 114; Ministry of Health, 114, 127, 129; National Institute of Human Nutrition, 115, 116, 119, 132; National Research Council, 114; nutrition advisor, 53; railway, 8
undernutrition. *See* marasmus (undernutrition)
UNICEF. *See* United Nations International Children's Emergency Fund (UNICEF)
United Kingdom, 133
United Nations International Children's Emergency Fund (UNICEF), xiii, 48, 49, 72, 113, 124, 139–40, 142
United Nations (UN), 49, 64, 136; Food and Agriculture Organization (FAO), 21, 43, 49, 66; General Assembly, 48; Protein Advisory Group (PAG), 65; Protein Board, 48
United States, xii, xiii, 16, 38, 65; President's Emergency Plan for AIDS Relief (PEPFAR), 129
United States President's Emergency Plan for AIDS Relief (PEPFAR), 129, 131, 133. *See also* acquired immunodeficiency syndrome (AIDS)
universities, 9
University College in London, 19–20
urbanization, 62, 67; mental, 63
Uringi, S.W., 114, 115
US National Academy of Sciences, 47; Committee on Protein Malnutrition, 47

vaccines. *See* immunizations
Vaughan, Megan, xiv
violence, xiii, 11, 23, 111, 112, 118, 119, 120, 122
vitamins, 16, 31, 137, 138, 140; deficiency, 14–15; research on, 143
vomiting, 57, 61, 62

Washington, DC, 126
weight, 18, 27, 92; charts, 92–93; deficit, 94; gain, 40, 57, 93, 94; loss, 57, 61
Welbourn, Hebe, 53, 55, 58, 59, 61, 82
Whitehead, Roger, 113–14, 116, 118, 119–20, 133
WHO. *See* World Health Organization (WHO)
Williams, Cicely, 14, 15, 20, 42, 49, 64
women's groups, 99, 107, 110
World Bank, 125, 126, 127, 130, 131, 133
World Food Program, 139
World Health Organization (WHO), xiii, 21, 23, 43, 44, 47, 49, 50, 51, 66, 129, 131, 137, 139; International Code of Marketing Breast Milk Substitutes, 66; Protein Advisory Group, 47–48
World War II, 7, 15, 18, 19, 20, 49

yaws, 12, 30–31